Ecology of
Soil Fungi

# Ecology of Soil Fungi

*David Michael*

## D. M. GRIFFIN

*Master, Burgmann College,*
*Australian National University,*
*Canberra, Australia*

Syracuse University Press

1972

*First published 1972*
*by Chapman and Hall Ltd*
*11 New Fetter Lane, London EC4P 4EE*


© *1972 D. M. Griffin*


*Published in the U.S.A. by*
*Syracuse University Press*
*Library of Congress*
*Catalog Card Number: 72-247*
ISBN *0-8156-5035-3*

*Printed in Great Britain by*
*Richard Clay (The Chaucer Press), Ltd*
*Bungay, Suffolk*

# Preface

This book has grown out of courses of lectures on the ecology of soil fungi given to final-year undergraduates in Agricultural Science at the University of Sydney, Australia, and, in 1969, to graduate students in Plant Pathology, Soil Science and related disciplines at Washington State University, Pullman, U.S.A. This background delimits the readers for whom this book is primarily intended – senior undergraduates and graduate students in the earlier years of their research. It is hoped that more experienced workers may also find some matter of interest within these pages.

In writing, I have deliberately retained the rather personal approach to the subject which is perhaps more characteristic of the lecture than the scientific article. Such an approach is congenial to me and is underpinned by a conviction that it is unbecoming to speak *ex cathedra* on a subject so vast and so fast developing as soil ecology.

The reader will perceive a progressive change in the nature of successive portions of the book. The first chapters are very general, others in the first part of the book only a little less so. The intention is there to provide a critical introduction to the background facts and concepts in the ecology of soil fungi and to provide an entry to the literature by means of a relatively small number of references, often to review articles. With this foundation laid, I have proceeded in Part 2 to treat a portion of the subject, physical ecology, in greater detail and with such rigour as is presently possible. More references will be found in this section, often to articles outside the normal literature of soil microbiology but relevant to the physical aspects of that subject. If the core can be at the end, Part 2 is the core and *raison d'être* of the book.

I should like to acknowledge my indebtedness to a number of my teachers and colleagues. From Professor E. J. H. Corner, F.R.S., I learnt as an undergraduate to respect the unconventional approach in things biological and to appreciate the importance of careful observation. Dr S. D. Garrett, F.R.S., introduced me to the whole topic of the ecology of fungi and has been a constant guide and friend. Where I have reached conclusions somewhat different to his, as expressed in his book *Pathogenic Root-Infecting Fungi*, I trust that he will consider a certain rebelliousness on the part of a student to be only right and proper in this day and age. Professor N. Collis-George and Dr D. E. Smiles of the Department of Soil Science, University of Sydney, have been my mentors in soil physics and much of my research has been greatly influenced by them.

This book was largely written, and nearly all my research work discussed in it was carried out, while I was a member of the Department of Agricultural Botany, University of Sydney. My indebtedness to the research students within that Department will be most obvious to them, for many of the ideas stated and positions adopted were formed in the light of their work and from discussion with them. Finally, but by no means least, I must thank my colleagues Professor N. H. White and Dr C. D. Blake for providing an atmosphere in which research was made easier. I am also grateful to Dr Blake for reading and criticizing the manuscript of this book.

D. M. GRIFFIN

*Department of Forestry,*
*Australian National University,*
*Canberra.*
*5th April 1971*

# Contents

## Part 2 · Physical ecology of soil fungi

# PART ONE   General Ecology of Soil Fungi

# 1 Introduction

Ecology is the 'study of the relations of animals and plants, particularly of animal and plant communities, to their surroundings, animate and inanimate' (Abercrombie, Hickman and Johnson, 1954). Here lies both the challenge and the danger of the subject. The challenge lies in the inherent complexity of all stages in the study. The first stage is the adequate description in factual terms of the biological and physico-chemical components of the system and of their changes in space and time. An informed consideration of this description suggests to the investigator the causal relationships between various components of the system and leads to controlled experiments, usually with simplified systems. Finally, the results of the various experiments and the hypotheses derived from them are integrated to form an explanation of the system revealed in the original description. The processes of ecological research are thus basically similar to those of all scientific research but in ecology the system studied is at its most complex and the integration is, ideally, the integration of a large component of all biological knowledge. Such a recital of the challenge of ecology immediately reveals the correlated dangers, for it is singularly easy to fall into error through a failure to describe accurately the various parts of the system and to appreciate their possible significance. An error at this stage may lead to the development of inapposite experimental techniques, so that the final synthesis must inevitably fail. A constant temptation besetting the ecologist is that of loose thinking to enable him to gloss over intracticable parts of his study.

The topic of this book is primarily the ecology of soil fungi. This branch of ecology presents both special difficulties and exceptional possibilities. There are two difficulties. The first derives from the unusually complex physical and chemical environment provided by the soil. Secondly, fungi are microorganisms (if one neglects the large sexual stages of certain species) so that the spatial scale is microscopic. A small volume of soil contains a vast assemblage of minute organisms of diverse form and physiology. Their small size, however, precludes the morphological complexity attained by higher plants and animals so that their reactions to each other and to their environment should be more easily attributable to basic physiological processes than is usually the case. The smallest and least differentiated free-living biological entities are the bacteria, but the ecology of these is hindered by the difficulties attending the first stage of investigation – description of the system. The lack of easily utilized specific criteria makes the accurate identification of bacteria and the precise description of their distributional patterns difficult to achieve. With the recent development of selective media for defined groups, this impediment may soon be less serious.

In fungi, the vegetative hypha approaches the simplicity and uniformity of structure of the bacterial cell but identification is more readily made by virtue of their differentiated reproductive stages. The bacterium, unicellular, small and with a large surface to volume ratio, is well-suited for the absorption of nutrients from a solution in which the cell is immersed. The hyphae of fungi and actinomycetes, however, permit the penetration of solid nutrient substrates by physical force and the extension of the mycelium from one substrate to another over nutritionally uncongenial areas. It is of interest that those fungi that resemble the bacteria in their morphological simplicity, notably the yeasts, are those that predominantly grow in situations where the nutrients are in the dissolved, rather than in the solid, form.

The ecology of soil fungi, therefore, has many attractive features: that many have found it so is shown by the size of the literature. Many who write on the topic would not consider themselves to be soil ecologists, but rather plant pathologists, for an ecological approach is a necessity when considering soil-borne diseases of plants. In this book, I have attempted to reduce to a minimum the degree of overlap with Dr J. L. Harley's *The Biology of Mycorrhiza* (1969) and Dr S. D. Garrett's *Pathogenic Root-Infecting Fungi* (1970) and the reader should refer to these books for recent treatments of topics that form a part of soil ecology in the wide sense. There is similarly a growing literature concerning the ecology in soil of fungal pathogens of man and animals and of those fungi which degrade materials as diverse as aviation kerosene and tents. The ecology of soil fungi is thus no ivory-tower pursuit but one with many applications to the ultimate welfare of mankind.

This book is unashamedly a personal view of fungal ecology. Such an approach I believe to be not only valid but necessary. The concepts and techniques associated with the ecology of soil microorganisms are now so complex and varied that no one person is competent to write authoritatively on them all. More than ever before, the soil biologist must have some precise knowledge of soil itself, for it is now clear that the activity of microorganisms and their interactions are profoundly affected by the physical and chemical framework afforded by the soil. The soil can no longer be looked upon as a vague background to biological activity: the two are intimately interconnected. In the past two or three decades, soil physicists have provided an understanding of the physical nature of soil which is adequate for the purposes of most biologists and in this area there is now the possibility of rapid advance.

## 1.1. The soil

We shall consider the soil to be the surface layer of the land, consisting of an intricate mixture of inorganic and organic components physically and chemically distinct from the underlying material. Most roots and soil organisms are located in this upper zone, the thickness of which is measured in inches rather than feet. Although it is convenient to consider the inorganic and organic components separately, it is fallacious to imply, as some workers have recently done, that the mineral soil is the 'true' soil and the organic matter (particularly in its identifiable form) an intrusion.

### 1.1.1. *Soil profile*

If a vertical section through soil is produced by digging a trench the soil profile is exposed. The top layers, largely organic in content, will be discussed later. The highest horizon that is predominantly mineral is designated $A$ and is eluvial (leached). It is frequently divided into two subhorizons. The $A_1$ horizon is distinguished by a relatively high organic matter content and usually dark colour. Beneath, the $A_2$ horizon has lost clay minerals, iron or aluminium, or even all three, so that there is a concentration of the more resistant minerals. Organic matter content is usually rather low. Cultivation destroys the top of the profile and a mixed horizon, sometimes designated $A_p$, results. Proceeding down the profile, the $B$ horizon is characteristically illuvial in that there is an accumulation of clay, iron, aluminium and/or organic matter.

B

Although roots and organisms penetrate into the $A_2$ and $B$ horizons, maximum biological activity is in the $A_1$ horizon.

In humid areas, leaching is intense and a marked $A$ horizon develops, frequently acid in reaction. In drier climates, however, leaching is reduced and the $A$ horizon may be shallow and near-neutral to alkaline. The $B$ horizons usually contain an accumulation of calcium carbonate and/or calcium sulphate, even though the parent rock may not be calcareous. The depth at which these calcium salts accumulate becomes shallower as the climate becomes more arid, and the depth of penetration of water is reduced.

Gilgais are areas of microrelief consisting of small basins, knolls or ridges in an otherwise level area and occur extensively on heavy soils in semi-arid areas. They are produced by alternate expansion and contraction of swelling clays during wet and dry seasons. This action lifts material from the $B$ horizon into the elevated portion of the gilgai and soil properties vary over small horizontal distances. Frost heaving is another mechanism whereby the soil profile is greatly disturbed.

### 1.1.2. *Inorganic matter*

The inorganic particles are derived, directly or indirectly, from parent rocks and their composition, especially that of the larger particles, reveals this derivation. Many of the rock minerals, however, are changed by the physical and chemical processes of weathering and most of the smaller particles are secondary in nature. A soil is thus a complex product of parent material, topography, climate, time and biological activity. Conspicuous among the secondary minerals are the colloidal clays, many of which are highly reactive and tend to dominate the chemical properties of the soil. The major mineral particle that occurs in its primary form in soil is quartz and most of the larger particle fractions consist of this unreactive silica.

Mechanical analysis of a soil is the procedure of breaking the soil aggregates down into the individual component particles and then determining the frequency of distribution of particles among various size fractions. This separation is performed by sieving (for the larger particles) and by sedimentation in a column of water (for the smaller particles). The velocity of sedimentation, by Stokes' equation, is proportional to the square of the diameter of spherical particles. The expressed results of both sieving and sedimentation thus assume that the soil particles are spherical and it needs to be borne in mind that the given diameters are equivalent diameters only. If the particles in reality have a plate-like form, the discrepancy can be considerable.

The sizes of soil particles have a continuous distribution: to divide them into fractions is arbitrary. This situation is reflected in the development of a number of different systems, although only two are in common use in the relevant literature. The more important fractions of these, the International and American systems, are given in Table 1.1.

Table 1.1. *Names and sizes of main soil fractions according to the International and American Systems*

| Name of fraction | Equivalent diameter of particles (mm) | |
|---|---|---|
| | International | American |
| Sand | 2·0–0·02 | 2·0–0·05 |
| Silt | 0·02–0·002 | 0·05–0·002 |
| Clay | 0·002 | 0·002 |

Once obtained, the mechanical analysis of a soil can be shown as a single point in a triangular representation. The surface of such a triangle is conventionally divided into areas characteristic of the soil textural classes, but two triangles are then necessary because of the different dimensional definitions of the fractions in the American and International systems: that for the American system is shown as Fig. 1.1. In the sense just described, the texture of a soil is thus a simplified description of the particle size distribution or mechanical analysis.

The term texture is also used in a different sense to express the field assessment of shear strength, that is the force necessary to move adjacent planes of soil over each other. The soil is worked between the fingers, to destroy its aggregates, at a water content such that the soil just sticks to the fingers. The force that must be exerted by the fingers to work the soil in this structureless state is noted: if it is great, the soil is a clay, if small, a sand. Soils of intermediary shear strength are loams. The terms 'clay', 'loam' and 'sand' are thus arbitrary divisions of a continuous scale of shear strength. Qualifications of these terms are applied if a particular fraction is apparent to the fingers: hence 'sandy loam', 'silty clay' (silt has a characteristic buttery feel). Additional guides to the textural classes are afforded by the fact that the cast formed by squeezing a moist sand is exceedingly fragile; that of a loam may be quite freely handled without breaking; and clay may even be rolled into a ribbon.

Clearly, the term 'texture' and the names of the textural classes

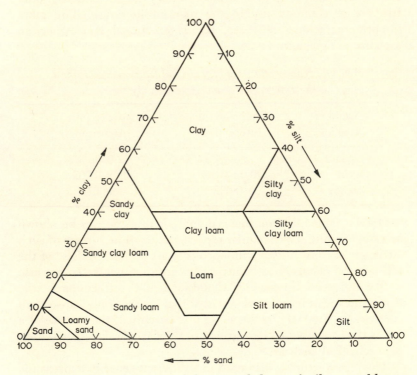

Fig. 1.1. *The composition of the textural classes of soils, as used by the United States Soil Survey*

are being used in two different ways. Usually the textural class obtained in one way will agree with that obtained in the other. Thus a soil determined as a 'sandy clay loam' in the field by its sandy feel and rather high shear strength will be shown to be a sandy clay loam by mechanical analysis. Occasionally, however, discrepancies occur so that the means of assessing texture should, strictly, be stated.

### 1.1.3. *Organic matter*

Above the $A_1$ horizon may be found organic horizons, the $A_{00}$ and $A_0$. In the biological rather than soil literature, this organic layer is frequently divided into three, the $L$, $F$ and $H$ layers. In the majority of soils, the surface layer consists of a relatively unaltered mass of identifiable plant remains. This is the $L$ (litter) layer or $A_{00}$ horizon. In a 'mor' soil the $F_1$ (fermentation) layer lies underneath the $L$ and consists of plant remains in the process of decomposition. The lowest organic layer is the $H$ (humus) layer in which the organic material is amorphous and no longer identifiable. The $F$ and $H$ layers together form the $A_0$ horizon. In a 'mull' soil the $F$ and $H$ layers are usually scarcely discernible because the activity of soil animals, particularly but not exclusively earthworms, leads to a rapid intermixing of the decomposing organic material with the surface mineral layer of the soil. In a 'mull' soil, the $L$ layer thus rests virtually directly on the $A_1$ horizon. Intermediate conditions between 'mull' and 'mor' are frequent and are sometimes referred to as 'moder'.

A characteristic feature of 'mor' is the slow rate of decomposition, a fact of considerable biological interest. It has been suggested that phenolic materials in the plant tissues tan the proteins and also form protective coatings over cellulose fibrils, thus rendering them relatively unavailable to microorganisms and the meiofauna. Such soils are also characterized by low base status so that they tend to be very acid.

Within the $A$ and $B$ horizons, organic matter can be divided into two components, that formed *in situ* and that derived from the $L$, $F$ and $H$ layers. Organic material formed *in situ* consists of the cadavers of soil animals, large or small, and the walls and cellular contents of bacteria, fungi and other microorganisms, but principally of the roots and other underground organs of plants. When dead, roots are clearly a part of the soil but they are not normally considered so when alive. A little reflection, however, will reveal the distinction to be tenuous because roots, by their excretions, greatly influence microbial activity within the rhizosphere. The rhizoplane flora, also, is largely saprophytic. The ecology of the rhizoplane and rhizosphere should thus be considered in any comprehensive treatment of soil ecology. I have chosen not to do so because these topics have been recently reviewed in a number of excellent articles (Schroth and Hildebrand, 1964; Viennot-Bourgin, 1964; Rovira, 1965a, b, 1969; Rovira and McDougall, 1967; Gams, 1967; Parkinson, 1967; Bowen and Rovira, 1968).

By the time organic matter from the aerial parts of plants becomes incorporated in the mineral layers of the soil, it has been greatly changed by a whole series of biological activities. (This is less so in an arable soil where cultivation incorporates relatively fresh material into the soil.) It is usually very comminuted, having passed through the intestines of earthworms or perhaps appearing in the faecal pellets of mites and springtails. Much of it is truly amorphous and may even be reprecipitated material. This humified organic matter may still contain some of the more resistant original polysaccharides (particularly some hemicelluloses) but in the main consists of mixtures of humic acid (soluble in alkali, precipitated by acids), fulvic acid (soluble in alkali, not precipitated by acids) and humin (not soluble in alkali). Although humic and fulvic acids and humin may make up 80–90% of the amorphous organic matter, their study has been difficult and rather unrewarding. In general, they have well-developed colloidal properties, swelling on wetting and with considerable base-exchange capacity, but their compositions and significance in the ecology of the soil is still imperfectly understood (Greenland, 1965a, b; Hurst and Burges, 1967).

The organic matter content of soils varies tremendously, from almost 100% in peat and 'muck' soils to less than 1% in impoverished soils. Perhaps a figure between 2 and 8% by weight would be typical of most agriculturally useful soils.

### 1.1.4. *Soil aggregates*

The individual particles of a soil are usually bound together into aggregates and the term 'structure' refers to the frequency, size, shape and stability of these aggregates. Soil structure profoundly affects the physical and chemical properties of a soil. In terms of structure, soils may be platy, prismatic, blocky or spheroidal and there are many subdivisions, for example, columnar, angular, granular.

Emerson (1959) has formulated a model of the structure of soil aggregates which is in accord with the known facts (Fig. 1.2). In the model, the clay crystals are grouped together into packs, or domains, or oriented platelets. Different domains may be held together by electrostatic attraction but in the main the binding substance is thought to be organic matter, binding quartz to quartz, quartz to clay domain and, often, domain to domain (Greenland, 1965*a*, *b*). Such a model has great microbiological implications, for the pores within an aggregate will usually be too small for bacteria and fungi to enter and much of the organic matter in a soil will be protected from degradation (Allison, 1968; Greenwood, 1968*a*).

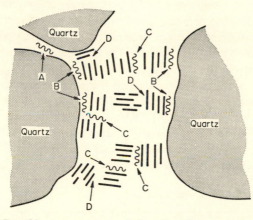

Fig. 1.2. *Possible arrangements of clay domains, organic matter, and quartz in a soil crumb. Different types of bond are indicated thus:*

    A. Quartz–organic matter–quartz
    B. Quartz–organic matter–domain
    C. Domain–organic matter–domain
    D. Domain (edge)–domain (face)

(After Emerson, 1959)

The soil pores are filled in various proportions with an aqueous solution and gases, of which oxygen, nitrogen and carbon dioxide are the most important. These pores and their contents are of great biological significance and they will be considered in detail in Part 2 of this book.

The reader requiring further knowledge on the genesis, morphology, classification and composition of soils should consult one of the numerous books on soil science, such as, Russell (1961), Leeper (1964) or the *Soil Survey Manual* (U.S. Dept. Agriculture, Handbook No. 18).

## 1.2. Soil organisms

We have considered briefly the inanimate soil and must now turn our attention to the great assemblage of organisms that live, or live partly, within it. Here the traditional divisions within biology have had a woeful effect. The zoologist, the microbiologist, the botanist, all study organisms or structures within the soil but precise information on interactions between major groups is almost totally lacking, except for those that fall in the field of plant pathology. Furthermore, a worker with one group has, in general, only the haziest appreciation of other groups and their ecology.

Our main interest is in the soil fungi, but before proceeding to consider them we shall note briefly and in general terms the place of other groups in the economy of the soil and their major relationships, if any, to fungi. More detailed information may be found in, or traced through, Burges and Raw (1967) for most groups of soil organisms, Waksman (1959), Alexander (1961) and Gray and Parkinson (1968) for bacteria and actinomycetes, Alexander (1961) for algae, Kevan (1962) for soil animals in general, and Sasser and Jenkins (1960) and Wallace (1963) for nematodes in particular.

### 1.2.1. *Organisms other than fungi*

*Bacteria.* Bacteria occur in counts varying between $10^6$ and $10^8$ cells per cubic centimetre of soil. Although unable to mechanically penetrate solid substrates because of their unicellular form, bacteria are extremely active under favourable environmental conditions because of their numbers and high surface area/volume ratio. They are largely responsible for the cycling and transformation within the soil of carbon, nitrogen, phosphorus, iron and sulphur. Consequently, the greater part of bacterial ecology up to the present time has a marked chemical orientation, and the more general aspects of the subject are little developed. Bacteria must interact strongly with fungi, both as they compete for nutrients, and sometimes oxygen, and by the production of inhibitory external metabolites. The precise investigation of the interactions between bacteria and fungi is likely to be a most rewarding field of research in the immediate future.

*Actinomycetes.* The most important genus of Actinomycetes in soil is *Streptomyces*, but its ecology has been gravely hampered by an inadequate classification and by technical difficulties. As late as

1969, it could still be debated whether streptomycetes occurred in soil mainly as vegetative mycelia or as spores, a topic largely elucidated for fungi at least a decade earlier. The conclusion appears to be that the conidial state is predominant, mycelium being largely associated with active growth at localized sites of nutrient concentration. Their high potential metabolic rate and proven ability as fungal antagonists, both as parasites and as producers of inhibitory metabolites, makes them of great significance to the mycologist.

*Protozoa.* Flagellates, amoebae, ciliates and testacea all occur in soil, and in arable and grassland areas there are populations of the order of $10^3$ $g^{-1}$ wet soil for each group. A high proportion of the population is usually encysted, particularly if the soil is not damp. The ecology of soil protozoa is still in its infancy but their prime importance must be as predators on bacteria, for this is the main food source. Predation is probably selective, although there is no certain evidence that protozoa are able to modify the bacterial composition of the soil. Fungi are no doubt occasionally utilized as food but this appears to be uncommon and direct interaction between fungi and protozoa is probably unimportant. Protozoan numbers are frequently correlated with root growth and thus with soil nutrient status. The increase is probably part of the rhizosphere effect, bacterial, and hence protozoan, reproduction being stimulated by root excretions.

*Nematodes.* The ecology of nematodes has advanced greatly in recent years and is probably in a comparable state to that of fungi, at least for the economically important pathogens of plants. Many feed on living protoplasm. This may be in the form of the cell contents of plants, including fungi, obtained by piercing the cell wall, or in the form of the whole protoplasm of bacteria, protozoa, etc., obtained by ingestion. While nematodes may thus feed on fungi, specialized fungi also trap, kill and colonize nematodes, but the quantitative significance of this activity is still in dispute. In contrast, the synergistic action of nematodes and fungi in initiating and maintaining infections in plant roots has been well established.

*Enchytraeid worms.* Enchytraeid worms (potworms) occur in greatest abundance in wet, cool, acid soils of high organic matter content. In such soils they may be abundant with populations of the order of $10^5$ m$^{-2}$. Soil enchytraeids are probably important agents in mixing plant debris with the mineral soil, but it is of interest to note that ingested plant material is little changed structurally by passage through the gut. This is in accord with data indicating that these worms lack the enzymes necessary to degrade the complex polysaccharides of plant cell walls. The food of enchytraeids has been little studied but probably consists largely of microorganisms, particularly bacteria and fungi. Potworms are unable to make their own tunnels in soil and so this has restricted them to existing voids.

*Lumbricid worms.* Because of their large size, lumbricid worms (earthworms) are the most obvious of the common soil inhabitants and their role in the economy of the soil has fascinated both professional and amateur biologists. Unlike most soil organisms, earthworms are able to move all but the largest soil particles. This is primarily done through ingestion of both mineral and organic matter, the latter being comminuted and partially decomposed during passage through the gut. It is not clear, however, whether the cellulases, chitinases and other degrading enzymes are produced by the worms themselves or by the bacteria and protozoa in their gut. The ingested material is finally ejected in the form of a casting, the physical and chemical properties of which differ greatly from that of the bulk of the parent material. In the present context, it is important to note that some earthworms ingest much organic matter lying on the soil surface and thus speed its incorporation into the soil. Casts are deposited either within the soil or at its surface. If the latter, a conspicuous stone-free surface layer gradually develops in natural soils but, in any case, a mixing of the soil ensues. The burrows tend to be stable and so an increase in the non-capillary pore space of the soil is another result of earthworm activity, with consequent effects on drainage, aeration and penetrability. Such a catalogue of effects suggests that earthworms might be of great significance in soil but the quantitative aspects of the topic need much further study.

Although earthworms probably affect soil fungi indirectly through their general activities, there is a direct effect of transportation (Hutchinson and Kamel, 1956) and recolonization of sterile soil by fungi from adjacent untreated soil is far more rapid in the presence of earthworms. It is not known what proportion of fungal units survive passage through the gut but there is no doubt that many spores do survive. Sporangia of *Phytophthora* spp. are also known to survive passage through snails which can thus act as vectors (Turner, 1967).

*Termites*. Although some termites have a localized mycological importance through cultivation of fungi in their nests as an apparently essential part of their diet, it would seem that their main significance must be as competitors with fungi for organic substrates. It has been estimated that the rate of utilization of organic matter by termites in one area of dry sclerophyll forest in Australia is approximately that of the total rate of input of organic matter into the soil, in the form of leaves, branches, trunks, etc. If such a situation is confirmed, its implications for fungal ecology are immense, for termites bring about a very thorough utilization of organic matter. What is not assimilated is concentrated in mounds and runways and consists largely of lignin and associated resistant material. This remaining organic matter becomes available for fungal growth only after many years, because the mounds are maintained for up to thirty years, and during this period, fungi are actively suppressed.

*Springtails and mites*. Springtails (Collembola) and oribatid mites (Acarina) are abundant in most soils. They are active in the breakdown of organic matter, both by comminution and partial digestion, and particularly favour material whose decay has already been initiated by microorganisms. There is thus almost certainly a strong interaction with fungi at the nutritional level. In addition, many springtails and mites feed on fungi, the gut content of some springtails in pine litter consisting of 64% hyphae and 21% spores (Poole, 1959). In a study of decomposition of bracken petioles, Frankland (1966) noted that fungi were heavily grazed by these arthropods and that some of the spiny species were often covered in fungal spores, probably aiding dispersal. Observation suggested that most of the hyphomycetes present on the petioles were palatable but that some species were preferred. Such selective grazing is to be anticipated because experiments have shown that fungi vary greatly in their ability to support populations of various mites (Sinha, 1964, 1966).

Other members of the soil meiofauna, such as millipedes (Elliott, 1970) ingest a considerable proportion of forest litter but their significance in the over-all process of decomposition is still poorly understood.

### 1.2.2. *Fungi*

In the first chapter, it was stated that the initial phase in ecology is one of description. With most macroscopic organisms, this phase is not one of experimentation but of direct observation and recording. Microorganisms, however, are usually invisible to the unaided eye and, in the case of soil fungi, are buried in an opaque matrix. Visualization must, therefore, precede description and this inevitably involves experimentation. To an exceptional degree, the basic data of fungal ecology, on which all else must rest, are only as good as the techniques used to produce a description of the pattern of fungal activity in time and space. Methods of study of the soil mycoflora have been ably reviewed in recent articles (Durbin, 1961; Menzies, 1963a; Warcup, 1965, 1967). Here, attention will be concentrated on problems of methodology rather than on a description of the individual techniques, many of which have been described by Johnson, Curl, Bond and Fribourg (1959).

The problems are well introduced by a consideration of a technique that was predominant in the earlier days of soil mycology and is still in frequent use. In this, the soil dilution plate method, a known weight of soil is shaken with a known volume of sterile water and a series of progressive dilutions is made. Sample volumes from each dilution are incorporated into just molten agar in a Petri dish and, after incubation, the number and identity of the resulting colonies is noted. As a result a description is obtained of the mycoflora of that soil in terms of the species present and their frequency per gram of soil. Such a description cannot be taken at its face value. The agar in a Petri dish on which the colonies have grown will, in normal practice, have a specific nutrient composition, a single $p$H, and a single temperature regime for incubation. Obviously, the fungi appearing abundantly on the Petri dish are those favoured, or at least not inhibited, by these conditions. Even were many different nutrient agars and many different incubation regimes to be used, the technique would still be selective and the resulting description biased. But what of the fungi with an inherently slow growth rate on all media; or those whose propagules are attached to coarse particles not incorporated in the dilution series because of their rapid sedimentation, or flotation on the liquid surface?

An even more fundamental difficulty involves quantification in soil mycology. What is the meaningful unit? Colonies derived from dilution plating arise predominantly from spores, so that the technique favours fungi that spore heavily in soil. Mycelia sterilia and basidiomycetes rarely appear. The frequency of colonies of a species in the Petri dish thus reflects, at best, the frequency of spores in the soil. But is the frequency of spores of major interest? Probably not. The frequency of colonies might at first sight be a better criterion until the indefiniteness of the fungal colony is considered. Objectively, the length of viable or better of actively growing hyphae of each species, along with the frequency of spores, are necessary to quantify the fungal population.

The soil dilution plate method clearly yields an extremely inadequate picture of the fungi in soil and is grossly selective even if the sole aim is to produce floristic lists. Much of the history of soil mycology is but an attempt to devise methods that would give a more adequate description. Like the dilution plate method, most are indirect and depend upon isolation of the fungi from soil. In the widely used soil plate method (Warcup, 1951), small quantities of soil are dispersed in agar in a Petri dish. The main advantage of the technique is that soil particles are not lost and genera such as *Pythium*, whose propagules appear to be associated with the coarser soil particles, are isolated. Fast growing fungi are still favoured, however, and the floristic list so produced does not greatly differ from that provided by dilution plating.

Much ingenuity has been shown in devising apparatus permitting the introduction of agar substrates into soil in the belief that their colonization will be primarily by those fungi whose hyphae were growing at the site of substrate introduction. The simplest and earliest device is the Rossi–Cholodny slide. A microscope slide is coated with a thin film of agar, buried and recovered after a set interval. The length of hyphae on the recovered slide was thought to be a measure of fungal activity at the site but the argument is weak, for the substrate is completely unnatural and the soil is inevitably disturbed by the slide's introduction. Isolation of the hyphae for subsequent colony identification is difficult.

In later devices, the agar has been protected from direct contact with soil by pouring it into perforated tubes or otherwise enclosing it within perforated boxes or plates (Chesters and Thornton, 1956; Anderson and Huber, 1965). Generally, the aim has been to isolate fungi active in the soil, rather than those existing as dormant spores. Although the latest techniques are probably fairly successful in this regard, the problems of selectivity of agars and of tardy or sensitive species are undiminished.

Apart from the multifarious indirect methods, there are a few which are relatively direct. Hyphae and fructifications may be observed in soil by examination through a microscope of natural or prepared soil surfaces (Kubiena, 1938; Casida, 1969). The system, however, leaves much to be desired optically and reveals only a minute proportion of the hyphae and spores present. Identification is possible only if reproductive structures are clearly visible. Soil may be sectioned after impregnation with plastic materials and the resulting sections examined microscopically for the presence of fungal structures (Burges and Nicholas, 1961). Identification is impossible and, in addition, it is usually impossible to distinguish between living and dead hyphae. The staining of prepared soil films (Jones and Mollison, 1948) is associated with similar difficulties although both techniques permit estimation of the prevalence of hyphae.

The implicit aim in much of soil mycology is to produce a complete list of the fungi present in the soil under investigation or, less commonly, of the fungi that are active at the time of investigation. I doubt, as I shall explain later, whether such knowledge is of much ecological significance and I doubt whether it is attainable. In many ways I believe such an aim to be a relic of those days when the soil was scarcely considered by mycologists or was vaguely conceptualized as a relatively uniform matrix within which fungi occurred. The old approach has, however, provided important basic knowledge, even though the information gained from any individual experiment has been largely determined and limited by the technique used (Warcup, 1957; Sewell, 1959c; Garren, 1964). In 1917, Waksman could ask 'Is there any fungus flora of the soil?' We now know the answer to be 'Yes'. Although strict delimitation is inherently impossible, there is an assemblage of fungi characteristically present in soil and, in the main, more abundant there than, for instance, on above-ground structures (Hudson, 1968) or in the sea (Johnson and Sparrow, 1961; Johnson, 1968). The number of fungal propagules within a soil sample is correlated positively with organic matter content (Jensen, 1934; Borut, 1960; Eicker, 1970) and so is highest in the surface soil and lowest in the *B* horizon (Warcup, 1951, 1957; Burges and Nicholas, 1961; Eicker, 1970). In general, species common in the surface soil are also common in the sub-surface, but some species appear to be proportionately more frequent in one or other of the layers. Such an appearance may be an artifact, for the species apparently commoner at depth are usually slow growing and so would be more difficult to isolate from the surface soil in the face of increased competition on the isolation medium (Warcup, 1951; Sewell, 1959*b*; Gray and Baxby, 1968; Eicker, 1970).

Peyronel and his associates at the University of Turin Botanical Gardens have made a valiant attempt to construct a system whereby the characteristics of the fungal population of a soil can be related to the environment (Peyronel, 1956). They divide the mycoflora taxonomically into eight groups (Phycomycetes; Melanconiales and Mycelia Sterilia; Tuberculariaceae and Stilbellaceae; *Aspergillus*; *Penicillium*; Dematiaceae; Ascomycetes and Sphaeropsidales; remaining Moniliales) and then note the number of species identified within each group. A high number of species within a group is considered to be characteristic of certain environmental factors, for example, Phycomycetes – cold, wet, high latitudes; Tuberculariaceae and Stilbellaceae – much organic matter; *Aspergillus* – hot, arid, low latitudes; *Penicillium* – less xerothermic than *Aspergillus*, higher latitudes.

The data are represented diagrammatically by drawing lines radiating at equal angles from a point. Each line represents a group and its length is proportional to the number of species recorded. The diagrams are considered to be both indicators of the taxonomic composition of the flora and of the dominant environmental factors, that is, to be indicators of mycological communities within the soil. We might question whether species are the relevant units, whether the groupings are logical and homogeneous, and Peyrenol has discussed many of these pertinent aspects. I believe, however, that there are two fundamental criticisms of any such scheme.

First, I doubt whether due weight has always been given to the difficulty of proving a negative, that a given species is truly absent in a given soil. In a study of the soil microfungi of willow-cotton-wood forests, Gochenaur and Whittingham (1967) showed that no less than 500 isolates from each of five sites were required to provide an adequate sampling, using the soil dilution technique, of the prevalent fungi capable of growing under the experimental conditions employed. Even then, no oomycetes or basidiomycetes were isolated. Few workers have taken so many samples, thus their lists must be suspect on this ground alone. As anyone who has conducted such research knows to his cost, a slight change in technique, of season of year, length of storage or some other unappreciated factor can determine whether a fungus can manifest itself and so be recorded. If such difficulties exist for a single worker, how much more careful must one be when comparing lists produced by different workers, often with different techniques. Using floristic lists compiled from different soils, it is possible to compare *recorded* fungi but only rarely to compare the fungi present *in* the soil. The process of description, of visualization, is as yet too uncertain for one to have confidence that, in general, the absence of a fungus from a list means it is not present in the soil. Confidence increases if special selective techniques are used to demonstrate presence or absence, but such techniques have rarely been used in the compilation of floristic lists.

The second criticism concerns the assumption that soil mycofloras will be clearly dominated by those environmental factors which act at the macroscopic, or even geographic, level. Geographical differentiation appears to be valid for parasites of above-ground organisms and for saprophytes colonizing leaves and stems (Pirozynski, 1968). In some cases, this is a direct effect of host distribution but in other cases differences may be attributable to the direct effect of a physical variable, such as temperature. In the sea (Johnson, 1968) or soil, the situation may well be different.

Another more quantitative approach to the delimitation of fungal communities has been adopted in a series of papers emanating from the Department of Botany, University of Wisconsin. Reference to all of these papers may be traced through Christensen (1969), who has succinctly summarized the aims and techniques of the group: '$A_1$ horizon samples, or the equivalent, have been used in all of the studies, similar collection and isolation procedures have been employed, and analyses have involved examination of 180–200 or more isolates from six, eight or ten sampling sites in each stand. The intent has been to describe a specific portion of the total fungal flora in individual communities, and by quantitative comparisons to describe the compositional variance among stands. The Wisconsin surveys have dealt only with that portion of the fungal population obtainable on soil dilution plates after a 3 to 7 day incubation period.' It is impossible to summarize in a few words all the results and conclusions embodied in this important series of papers, and I hope I cause no distortion by selecting the following points as particularly noteworthy.

A few fungi appear to be ubiquitous in Wisconsin soils but the fungal communities of the three major vegetational units (prairies, southern upland forests and northern upland forests) are generally dissimilar. Dissimilarity lies both in species composition (Table 1.2) and in the relative abundance of a given species. Some

Table 1.2. *Measure of dissimilarity in the species present in the soils of three major vegetation units (after Christensen, 1969)*

| Vegetational unit from which species were isolated | Percentage of species found only in unit of origin | Percentage of species found in other units | | |
|---|---|---|---|---|
| | | Prairies | Southern upland forest | Northern upland forest |
| Prairies | 70 | — | 25 | 14 |
| Southern upland forest | 53 | 22 | — | 24 |
| Northern upland forest | 71 | 15 | 29 | — |

species, notably of *Fusarium*, were common in the prairie soils but were missing or rare in the forest soils, a pattern supported by work in other parts of the world (Thornton, 1960; Park, 1963).

The majority of species from the prairies were found at all sampling sites within that one major vegetational unit, but in the northern conifer-hardwood forests, 90% of species were isolated from fewer than four of the thirty-six forest sites sampled. In the latter case, therefore, there were fewer indicator species characteristic of the vegetational unit even though there was no evidence of discrete communities within the unit. The greater abundance of narrow amplitude species in the northern forests than in the prairies is probably indicative of a greater environmental diversity in the former. Evidence was obtained that climate, vegetation and

C

edaphic factors were all effective in limiting the distribution or frequency of various species. It was further concluded that 'antibiotic interaction may have as much to do with species composition in soil microfungal communities as have vegetation and related physical–chemical aspects of the soil environment'.

Although the Wisconsin workers believe that they can discern fungal populations characteristic of vegetational–edaphic units, this has only been as the result of the sophisticated analysis of a large body of data and they are careful to apply geographical limits to their claims. They state that 'most of the species isolated in the Wisconsin studies are world-wide in distribution and can be thought of as "hemlock" or "wet prairie" species only within prescribed geographic limits'. Elsewhere a combination of quite different environmental factors may create another favourable situation resulting in the presence, and even abundance, of a species.

Domsch, Gams and Weber (1968) conducted an intensive survey, involving 25,000 isolates, of the soil fungi of two fields in Germany. Both were originally wheatfields but sections were given over to the monoculture of wheat, peas and rape during the experiment. The quantitative distribution of twenty-eight species of fungi was significantly different under the three monocultures, and the data from areas cropped continuously with wheat were similar to those from three Dutch wheatfields. Elsewhere, in South Africa, U.S.A. and Australia, *Penicillium urticae* has been shown to be a characteristic component of the mycoflora of wheat fields (Jooste, 1966; Norstadt and McCalla, 1969; personal unpublished data). Such results indicate an influence of the crop on the soil mycoflora, but it is important to emphasize that differences are essentially quantitative: significant differences on a simple present or absent basis are few.

In general terms, the mycofloras of different soils, even of greatly different geographical areas, are striking for their similarity rather than dissimilarity. There is no obvious correlation between geographical area and great assemblages of species such as occurs in higher plants. Thus the concept of biomes, applicable to higher organisms, has little relevance to soil fungi. There are probably no species, or even genera, common to the higher plant floras of the humid, tropical Congo basin, the arid, tropical savannah of Somalia and the cold, humid Italian alps. Yet, a perusal of six mycological articles concerning these areas reveals fifteen common fungal genera and the following species are among those recorded from them all: *Aspergillus niger, A. ruber, A. versicolor, Chaetomium globosum, Fusarium oxysporum, Gliomastix convoluta, Memnoniella echinata, Penicillium rubrum, Stachybotrys atra* and *Trichoderma viride (sensu lato)*. This well-nigh universal distribution of many microorganisms has been explained by the omnipresence of specific microenvironments suited to the various species (Stanier, 1953). If an ecological niche is 'a substrate, existing within an environment, over a period of time' (Tribe and Williams, 1967),

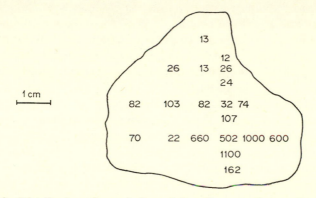

Fig. 1.3. *Distribution of population of fungal propagules on the freshly exposed surface of a fractured clod of soil. The figures represent colonies per mg of soil isolated from the indicated points.* (After Dobbs and Hinson, 1960)

then the key to understanding seems to be the realization that the scales of time and distance applicable to microorganisms are extremely small and that soil is extremely heterogeneous on these scales.

Fig. 1.3 shows that the variation in magnitude of the spore population found at points within a single clod of soil, 6 cm across, can be of the order of one hundredfold, a difference probably based on differences in nutrient concentration and pore size. In particular, the uneven distribution of some plant pathogens can be largely attributed to the production of propagules within, or immediately adjacent to, host roots (Trujillo and Snyder, 1963). Studies conducted in England (Williams and Parkinson, 1964; Parkinson and Balasooriya, 1969) and Germany (Gams and Domsch, 1969) have shown that the spatial variation in percentage of soil particles colonized by a range of fungi was sufficient to mask any seasonal variation but such is not the case in less temperate climates (Saksena, 1955; Warcup, 1957; Moubasher and El-Dohlob, 1970). Indeed, at one English site, the more sensitive technique of soil-sectioning showed more mycelium in the soil in autumn and winter, coinciding with the period of maximum leaf fall and hence nutrient input (Nicholas, Parkinson and Burges, 1965). Sewell (1959c), however, has shown that the mycelial activity of *Trichoderma viride* is reduced in winter although the reduction is not revealed by dilution plating where, as we have seen, colonies are normally derived from conidia.

The power of dormant survival of microorganisms, mainly as spores, is important, for by dormancy they are able virtually to ignore periods that are inclement for physical or chemical reasons. These may together cover much the greater part of each year, but the few days when conditions are congenial permit them to maintain their populations. The ephemeral annual of semi-arid areas is analogous but the brevity of its activity makes it unusual among higher plants, whereas such bursts of activity in the midst of predominant inactivity appear to be the norm for microorganisms (Warcup, 1957). There is little information on the perennial, consistent growth of fungal colonies, but it probably exists in those basidiomycetes utilizing large volumes of comparatively uniform and resistant organic matter. Outstanding examples are the 'fairy-ring' fungi of permanent grasslands and the fungi causing heart-wood decay in trees, both of which can persist for decades. The activity of most fungal colonies in soil, however, is measured in days or, at most, a few weeks.

An adequate description of the soil mycoflora must take account of both the heterogeneity of the matrix in space and time and the varied morphological, physiological and biochemical attributes of the fungi themselves. If this be accepted, the techniques applicable to the task of describing the composition and activity of soil fungi will be quite different from those akin to dilution plating. A simple and obvious modification in approach is to enumerate the fungi associated with the various component fractions of soil, mineral and organic, large and small particles. To this end, a number of techniques involving soil partition as a first stage have been developed (Parkinson and Williams, 1961; Gams and Domsch,

1967; Hornby, 1969). Such techniques have revealed, for instance, the presence within organic fragments of the hyphae of *Ophiobolus graminis* and *Rhizoctonia solani*, the oospores of *Pythium* and *Aphanomyces* and the chlamydospores of *Fusarium* spp. Sterile mycelia, many probably of basidiomycetes, have been isolated from the surface of the heavier mineral particles and soil sieving has yielded directly such larger bodies as the zygocarps of *Endogone* and sclerotia (Gerdemann and Nicholson, 1963). Care, however, needs to be exercised in the interpretation of the results of experiments using soil fractionation because it is incorrect to think of the mineral soil fractions as being spatially separate within the soil (Emerson, 1959). In most soils, sand, silt and clay will be intimately mixed within aggregates and the apparent association of hyphae or other structures with one or more fractions is very likely to be an artifact. The frequent association of hyphae with large mineral particles may merely reflect a relative abundance of crevices and other sites providing protection from the effects of washing and sieving. Papavizas (1968) has also pointed out that hyphae are most unlikely to pass through fine sieves because of their length and therefore will be rarely recorded from silt and clay fractions even though they may have initially been associated with them. If persistence and subsequent saprophytic ability of sclerotia increases with size, viable propagules will inevitably appear to be associated with large soil particles if distribution is determined by sieving.

Great quantities of cellulose are incorporated annually into the soil from vegetation but its utilization is not easily studied in unaltered soils. The study, however, is greatly aided if the cellulose content of the experimental soil is artificially augmented by the addition of squares of cellophane. Such squares can be buried and recovered at various intervals of time and the fungi colonizing them identified *in situ* or by isolation. The activity of the various species can be assessed in terms of area of cellulose destroyed, for this can easily be seen in the transparent cellophane, and successions can be established. The basic idea of the technique can obviously be employed in the investigation of many soil environments of different organic status and studies of the colonization of plant roots, shells of groundnuts, hair and even buried cadavers of pigs come to mind. We shall consider some in more detail later.

The approach outlined above lays emphasis on the substrate. An alternative approach is to lay emphasis upon a particular organism or group of organisms. The utility of selective techniques can be demonstrated with three examples. Until recently, the Gymnoascaceae were a relatively little-known group within the Ascomycetes and virtually nothing was known of their distribution or ecology. The majority of members are keratinolytic and this led to the development of a technique of enriching the soil with hair or other keratinaceous substrates so as to provide a bait for these fungi. The technique has proved amazingly successful and has resulted in the isolation and description of many new genera and species (Apinis, 1964; Somerville and Marples, 1967). It has become obvious that the Gymnoascaceae are a characteristic component of the mycoflora of most soils yet are rarely isolated by any but the selective enrichment method. This discovery is important because some members of the Gymnoascaceae, and supposedly-related imperfect species (Kuehn, Orr and Ghosh, 1964), are causes of dermatomycoses in animals, including man, and it is apparent that the soil is a reservoir for a number of pathogens in the genus *Nannizzia* and *Arthroderma*.

A second example of the successful use of selective techniques concerns *Amorphotheca resinae*. The fungus first came into prominence when its imperfect stage, *Cladosporium resinae*, was isolated from creosoted telegraph poles and railway sleepers. More recently, the prime economic importance of the species has resulted from its growth on aviation kerosene, leading to damage to aircraft fuel systems (Hendley, 1964). Research has shown that strains of *A. resinae* are unusual in being able to utilize a number of hydrocarbons, vegetable oils, waxes, resins and other substrates, and in being remarkably tolerant of substances, such as creosote and terpenes, which are normally toxic. Strains can also break down agar, yet some are unable to utilize far simpler carbon sources such as sucrose and cellulose. The species had not been isolated from soil until Parbery (1969) placed freshly creosoted matchsticks on to soil samples. *C. resinae* then grew in almost pure culture on the matchsticks and surfaces of many soils. The success of the technique appears to lie more in the suppression of antagonists by the creosote rather than in any directly stimulatory action of that material. Parbery has shown the species to be present in the mineral, but not in the litter, horizons of many soils and has suggested that the fungus there grows on natural hydrocarbons, waxes, steroids and terpenes.

Thermophilic fungi are those that have a maximum temperature for growth at or above 50°C and a minimum temperature for growth at or above 20°C. Earlier surveys showed such fungi to be quite rare in soil. Recently, however, ten of the known seventeen species (along with a thermophilic streptomycete and two thermotolerant fungi) were isolated from a single English pasture soil by the use of selective techniques (Eggins and Malik, 1969). Many of these fungi are cellulolytic and the technique best demonstrating their presence reflected this. Soil was amended by the addition of cellulose powder and then incubated at 50°C for 12 days. Soil samples were taken at intervals and were incorporated into agar containing cellulose at the sole carbon source. The complete technique thus utilized an initial enrichment culture, a selective temperature and a final selective isolation medium. Slight variations in the technique were found to favour individual thermophiles but the principle remained the same.

Selective techniques have also proved valuable in plant and human pathology, baits used including carrot discs for *Thielaviopsis basicola*, pineapple leaves for *Phytophthora cinnamomi*, and dehydrated pork liver cubes for *Coccidioides immitis*. If such techniques can be combined with the 'most probable number' method of biometrical analysis, accurate assessments of populations can be made, as has been done for *T. basicola* (Tsao and Canetta, 1964). Recent developments in the chemical industry have led to the production of many chemicals that retard the growth of many microorganisms, while permitting the growth of a few. Nutrient agars containing one or more of these chemicals provide selective media of great use in isolating specific organisms or groups from soil or other material (Domsch and Schwinn, 1965; Tsao, 1970).

Another new technique of great promise for the study of individual species in soil is provided by immunofluorescent staining. Specific antisera for the antibodies of the microorganism are prepared and used to produce fluorescein isothiocynate γ-globulin conjugates. Soil samples containing the microorganism are then placed in staining troughs containing solutions of the conjugate and are subsequently examined microsopically under ultra-violet light. The species for which the antiserum was prepared then fluoresces brightly. In the absence of cross-reactions which, on limited evidence, are not important for fungi, the technique is specific. It has been used for bacteria (Hill and Gray, 1967) and fungi (Eren and Pramer, 1966) in soil and is being developed for the study of plant pathogens in the rhizosphere. Fluorescent brighteners are also being used to study the behaviour of marked spores and mycelia in soil (Grossbard, 1958; Eren and Pramer, 1968) but the technique is limited by dilution accompanying growth.

We have noted above that the soil abounds in microsites differing in space and time in many chemical and physical attributes. Similarly, we have noted that the fungal flora of the soil is diverse in morphology, physiology and biochemistry. The remainder of this book will be concerned with studying how the interactions between the inanimate soil, the fungi and the rest of the living organisms leads to the co-existence of hundreds of different microorganisms within very small spatial limits.

# 2 The significance of substrate composition

## 2.1. Nutrition of fungi

### 2.1.1. *Carbon*

Fungi are heterotrophic in that they are able to synthesize organic material sufficient for growth only from preformed organic components. (Many fungi can assimilate carbon dioxide, but the range of products, mainly organic acids, is limited.) They are also chemotrophic, in that they obtain energy independently of light by chemical reactions. The energy-yielding reactions of fungi are based on organic substances and respiration is typically aerobic, little growth occurring in the absence of oxygen as a terminal electron acceptor. Fungi are thus doubly dependent upon organic matter, and competition for the more easily utilized forms will often be intense.

With a few exceptions, fungi in culture are able to utilize the commoner hexoses and disaccharides but progressively fewer utilize potential carbon sources as they become increasingly complex (Cochrane, 1958). Such a situation is easily explicable on biochemical grounds, for it is likely that carbon nutrition is initiated by the entry of one of the phosphorylated derivatives of certain trioses, pentoses or hexoses into the respiratory pathways and in this connection glucose is pre-eminent (Blumenthal, 1965). Utilization of a potential carbon source by a fungus therefore implies the ability to produce phosphorylated derivatives from that source. Phosphorylated derivates may appear directly as products of cleavage reactions; if not, the organism must possess the specific kinase permitting phosphorylation of the simple derivatives themselves.

For a carbohydrate in the external medium to act as a respiratory substrate, it (or its breakdown products) must be able to pass through the cell wall and the cell membrane. Entrance of sugars is effected by carrier mechanisms (sometimes called permeases) specific to each sugar, or possibly specific to groups of sugars with the same molecular shape. Thus *Saccharomyces cerevisiae* possesses a carrier for maltose which is thus able to enter the cell where the α-linkage is cleaved by maltase to yield two glucose molecules that are subsequently phosphorylated. *S. cerevisiae* lacks, however, a carrier for sucrose and this sugar is cleaved extracellularly to glucose and fructose by invertase located on the cell surface. The monosaccharides are then transported into the cell by specific carriers. Organisms may possess different permeases for the same substrate. Thus the maltose permease of *S. cerevisiae* functions both aerobically and anaerobically whereas that of *Mucor rouxii* is dependent upon a functional respiratory chain (Flores-Carreon, Reyes and Ruiz-Herrera, 1969) and such differences may yet be found to be reflected in ecology. Carriers are sometimes inductive systems, being formed only in the presence of the specific sugar group and inhibition can also occur.

The greater quantity of respiratory substrates occurring naturally in soil are, or are derived from, the cell walls of plants which consist primarily of complexes of sugars or sugar-containing molecules (Gupta, 1967). The degradation of these complexes involves an extraordinary number of enzymes (Mandels and Reese, 1965; Bateman and Millar, 1966). In a recent review (Albersheim, Jones and English, 1969) it has been stated that 'There is in all biology no molecular interaction more specific, more sensitive to structural alteration, than the recognition of saccharides by proteins.' Again, 'an exo-β-1,3-glucanase, for example, has little ability to degrade β-1,4- or β-1,6-glucans, while a β-1,4-glucanase does not hydrolyze α-1,4-glucans or β-1,4-galactans. These specificities reflect the monosaccharides connected by the glycosidic linkage, as well as both the configuration (α or β) of, and the carbon atom involved in, the glycosidic bond. In recognizing the components of a polysaccharide, degradative enzymes respond to the presence of such substituents as an amino group at carbon atom 2, a carboxyl group at carbon atom 6, methyl ether groups, and esters of both hydroxyl and carboxyl groups, as well as to the stereochemistry and oxidation state of a particular carbon atom. Such specificities as these imply that minute alterations in the structure of a polysaccharide can dramatically affect its susceptibility to degradation by a particular enzyme. Since polysaccharides appear to be composed of repeating units or to contain recurring sugar linkages a minor structural change could be so magnified as to become a major factor in determining a polymer's susceptibility to degradation.' The ecological implications of these statements are obvious for it is unlikely that a given fungal colony will produce more than a few of the enzymes necessary to degrade the many carbohydrates present in a natural substrate.

Carbohydrate utilization is further complicated by the phenomena of induction, inhibition and repression. (Zalokar, 1965; Mandels and Reese, 1965.) Some enzymes, like carrier systems, are produced only in the presence of the substrate and are then termed 'inductive'. Sometimes they are not produced even then unless the medium contains a small 'priming' quantity of readily utilizable hexose and this is most frequently so if the inoculum consists of small spores. The probable explanation is that production of the induced enzyme is itself an energy utilizing process that cannot be maintained by the meagre energy supply of the spore. If, however, the small quantity of hexose is assimilated, sufficient energy becomes available for enzyme induction and assimilation of the major carbohydrate source. Cellulase synthesis in *Pyrenochaeta terrestris* is subject to induction but it is also repressed by glucose in concentrations above 0·0005 molar. If degradation of cellulose occurs at a rate in excess of the ability of the fungus to utilize the resultant glucose, repression occurs and the rate of cellulose degradation falls until a balance of glucose production and utilization is attained. Failure to produce an enzyme by a fungus that does produce it under other circumstances is common, but in many cases the biochemical mechanism is unknown.

Carbohydrate utilization, therefore, can be blocked at many points. The organism may not produce the enzyme necessary to cleave the relevant glycoside linkage, induction may fail, or repression or inhibition may occur. Entrance of simple carbohydrates may not occur because of the lack of the relevant carrier system, or it may not be induced or it may be inhibited. Even if the carbohydrate enters the cytoplasm, it may not be utilized because of the absence of the necessary phosphorylating enzyme.

### 2.1.2. *Nitrogen*

Although assimilable amino acids and related products are present in small quantities in soil, particularly in the rhizosphere, most nitrogen available for uptake is in the form of ammonia or nitrate. Most fungi are able to utilize both ammonia and nitrate although ammonia is preferred (Nicholas, 1965). Many soil fungi, however, appear to need amino acids or yeast extract in order to grow at their maximum rate (Atkinson and Robinson, 1955). Many wood-destroying basidiomycetes, too, fail to use nitrate but grow luxuriantly on amino sources. Such a requirement may be associated with their growth on massive plant remains, where much of the nitrogen is in a complex form. Nitrate utilization is dependent upon the induction of nitrate reductase and is inhibited by ammonia so that in systems containing both, ammonia is largely utilized before nitrate. The inductive nitrate reductase is not formed by some fungi which are then dependent on ammonia or amino acids. Utilization of ammonia and nitrate is also dependent upon hydrogen ion concentration for in unbuffered systems, up-take, particularly of ammonia, causes rapid change in *p*H and this can confuse results of *in vitro* experiments. Apparent requirements for organic acids when ammonia is the sole nitrogen source are primarily the result of the buffering action of the acids. Such effects are unlikely to be of importance in most soils, because soils are intrinsically well-buffered.

### 2.1.3. *Carbon : nitrogen ratio*

Assimilable nitrogen can sometimes be a limiting factor to micro-bial activity but this has rarely been demonstrated except in the presence of abundant carbohydrates (Stotzky and Norman, 1961*a*; Finstein and Alexander, 1962). As it is thus the amount of nitrogen in respect to an amount of carbohydrate that is of importance, the matter is most usually discussed in terms of the carbon:nitrogen (C:N) ratio. The majority of fungi and bacteria have themselves C:N ratios of about 12:1 to 10:1 whereas most fresh plant remains have far greater ratios. In the initial stages of growth on such sub-strates, all the nitrogen becomes immobilized by incorporation into new microbial cells, but some of the carbon is lost through respiration as carbon dioxide. The C:N ratio is thus progressively reduced until, between ratios of 30:1 and 15:1, the rates of im-mobilization and mineralization are about equal. With ratios below 15:1, nitrogen will be mineralized and made available to higher plants (Scarsbrook, 1965). Over long periods, the C:N ratio of the decomposed organic matter will approach that of the micro-

organisms themselves. Should the initial substrate have a low
C:N ratio, say 8:1, then nitrogen mineralization will occur from
the beginning and the ratio will gradually increase.

The above general remarks on nitrogen nutrition and the significance of the C:N ratio need qualification in the light of recent
work on the growth of fungi on substrates extremely deficient in
nitrogen (Merrill and Cowling, 1966; Levi, Merrill and Cowling,
1968; Levi and Cowling, 1969). The mycelium of fungi grown on
nutrient agars routinely used for the cultivation of fungi contains
about 5% nitrogen by dry weight and has a C:N ratio of about
10:1. The C:N ratio of heartwoods, however, varies from 500:1
to in excess of 1,250:1 and, on natural and artificial substrates with
such ratios, some fungi adapt their mycelial nitrogen content to
that of the substrate. The nitrogen content may then fall to as low
as 0·2%.

Growth under these exacting conditions implies great efficiency
in nitrogen metabolism and a number of factors are probably involved. Mycelial nitrogen is re-used through internal translocation
of cytoplasmic constituents, or their autolytic products, from old to
young cells. Extracellular lytic enzymes are released that render the
cell wall and other constituents available for assimilation by younger
cells. Such a re-use is indicated by the ability of a number of wood-
inhabiting and wood-destroying fungi to utilize various fractions of
their own mycelium as a sole nitrogen source so long as glucose
was available as a carbon source. Proteins, peptides and amino
acids supported more rapid growth than nucleic acids, nucleotides
or cell wall materials.

A second factor involved in the efficiency of nitrogen utilization
is the preferential allocation of available nitrogen to metabolically
active systems that are highly efficient in the utilization of wood
constituents. Thus a change of C:N ratio from 25:1 to 1,600:1
increased the proportion of nitrogen present in the nucleic acid
fractions of *Polyporus versicolor* from 4·1% to 25%. This was
accompanied by a shift from the synthesis of amino acids, peptides
and proteins to that of lipids.

Although some fungi can re-use nitrogen, they only do so in the
presence of exogenous carbohydrates. In wood, cellulose is the
predominant form and it is of interest that cellulolysis diminishes
with increasing C:N ratios. The effect is not uniform among the
various species, however, and fungi capable of causing white rots

of wood are unique in their ability to produce diffusable cellulolytic enzymes at a C:N ratio of 2,000:1. Production of such enzymes by many other cellulolytic fungi is greatly reduced even at a ratio of 200:1.

Differential effects associated with nitrogen metabolism are clearly important in studies of wood degradation and no doubt will be shown to be so in the litter layer. A particular aspect warranting investigation is the significance of differences in nitrogen metabolism in determining the length of saprophytic survival in plant residues of high C:N ratios (see Chapter 4).

### 2.1.4. *Minor nutrients*

Apart from sources of organic carbon and mineral nitrogen, fungi also require for growth the usual series of minerals and trace elements as do higher plants (Cochrane, 1958). In addition, many require complex growth factors (Atkinson and Robinson, 1955; Fries, 1965). Little is known of the influence of these requirements of fungi in soil but they may be supposed to be minimal because such needs are likely to be supplied by all but quite exceptional soils. Inadequate phosphorus and sulphur have, however, been shown to limit respiration in a poor soil to which glucose had been added (Stotzky and Norman, 1961a, b).

### 2.1.5. *Physical form of the substrate*

The chemical nature of a substrate will obviously determine in part the ability of a microorganism to assimilate it. A less appreciated determining factor is the physical form of the substrate. Cellulolytic fungi growing on cotton fibres within the boll wind around the fibres before they penetrate the secondary wall (Simpson and Marsh, 1969). This superficial helical growth closely parallels the helical structure of the secondary wall of the fibre and it is known that both cellulolytic bacteria and the enzyme cellulase *in vitro* attack fibres in a similar helical fashion. It is to be presumed that this pattern of activity reflects differential susceptibility of the cellulose to lysis, probably associated with weaker bonding between cellulose molecules of adjacent helical units than between those in the same unit. Massive, homogeneous structures are thus likely to be more slowly degraded than those formed of sub-units with associated lines of weakness (Mandels and Reese, 1965). Such reasoning may explain the inability of the bacterium *Cellvibrio* to attack regenerated cellulose in the form of cellophane sheet although it is able to utilize the fibrous cellulose in filter papers (Keynan, Henis and Keller, 1961).

English (1963, 1965, 1968) has studied the growth of fungi on cellophane and on keratinized substrates. Despite its appearance, cellophane is not physically homogeneous but consists of many sheets of molecules lying in the plane of the film. Strong bonds exist between molecules in the same sheet but only relatively weak hydrogen bonds link the primary hydroxyl groups of the peripheral molecules of adjacent sheets. In hard keratinized substrates, such as nails and the cortex of hair, keratin (a complex protein with disulphide bonds) is deposited within animal cells until the cell becomes filled. The crystalline keratin units are thus separated by a periphery of cellular debris and cement, consisting of polysaccharides, mucopolysaccharides and proteins. Few organisms possess keratinases but many are able to degrade the intercellular material. Both cellophane and keratinized substrates, in quite different ways, are thus characterized by zones of easier availability which are of considerable length in two dimensions but very small in the third. Benefit will thus accrue to an organism that can penetrate the resistant zones and exploit the planes of weakness between them.

English has shown that if attack on such substrates is not to be localized to the exterior, two attributes are important. First is the ability of the fungus to grow along such very thin planes of weakness. Such growth is usually in the form of frond-like mycelium in which the cells are massed together, irregular in shape and of similar length and breadth but greatly diminished height. Second is the ability to produce 'boring' hyphae that penetrate mechanically the intracticable portions of the substrate. Thus species of *Curvularia* penetrate the dense keratin of hair cortex by very fine hyphae which then produce frond-like mycelium along the intercellular planes of weakness. All the evidence indicates that the boring hyphae penetrate by mechanical pressure and that enzymatic weakening is unimportant. There is thus a division of function between cylindrical hyphae that penetrate and frond-like mycelium that absorbs nutrients. Although the distinction is clear-cut in *Curvularia*, it is less so in keratinophilic fungi growing on hair *in vitro*. With them, the boring hypha often becomes modified into a multi-cellular perforating organ which undoubtedly combines enzymatic lysis of the keratin with mechanical penetration. Boring hyphae and frond-like eroding mycelium can also be demonstrated in many fungi growing on cellophane, but they are not produced by the same fungi growing in amorphous cellulose dispersed in agar. The form of the fungus is thus a response to the physical configuration of the substrate. This effect of the substrate has been little investigated as a factor in assimilation and ecology but English's work suggests its importance.

### 2.1.6. *Comparison of fungal and bacterial nutrition*

As bacteria are the main competitors with fungi for soil nutrients, it is appropriate to note certain differences between their responses to nutrient substrates. Thus fungi grow fastest on media of relatively high C:N ratio whereas the converse is true for bacteria. (The C:N ratios for potato-dextrose agar and nutrient broth are 180:1 and 10:1 respectively (Levi and Cowling, 1969).) Furthermore, fungi in general appear to exceed bacteria in the efficiency of their carbohydrate utilization, 30–50% of substrate carbon being transformed into new fungal cells. Fungi thus tend to immobilize nitrogen more strongly than bacteria and mineralize it less rapidly from a complex substrate.

Initiation of growth of colonies of *Escherichia coli*, *Klebsiella aerogenes* and *Streptococcus faecalis* occurred only on media containing at least 0·09, 0·013 and 0·005 g glucose $l^{-1}$ respectively, whereas *Aspergillus nidulans*, *Penicillium chrysogenum* and *Mucor hiemalis* all grew rapidly on media containing 0·009 g glucose $l^{-1}$ (Pirt, 1967; Trinci, 1969). Indeed, the maximal growth rate of *M. hiemalis* was attained at only 0·075 g glucose $l^{-1}$ whereas *E. coli* required 4·0 g $l^{-1}$. Trinci has also noted fundamental differences which are of profound ecological importance between the growth of bacteria and fungi. He has written: 'The filamentous habit permits variation in colony density such that the energy source available in the medium is used to maximum efficiency in extending the colony's diameter. The hyphal density control mechanism is presumably of significant selective advantage as it enables moulds to spread across substrates containing low concentrations of nutrients at near maximum radial growth rates; when a substrate containing a high concentration of nutrients is encountered, the colony grows more densely, produces aerial hyphae and may sporulate. Unicellular microorganisms do not possess a specific mechanism to control the density of growth within the colony and typically form small colonies. A further advantage imparted by the filamentous habit is that it facilitates penetration of solid media and as a result the organism absorbs nutrients which usually only become available to unicellular microorganisms by diffusion.' We shall have cause to return to this issue in Part 2.

## 2.2. Nutrition and substrate colonization

### 2.2.1. *Successions*

The breakdown of natural products by the mixed soil flora and fauna is such a complex process that it is impossible to discern with accuracy all the factors determining the sequence of observed events. Tribe (1960), however, studied the colonization and breakdown of cellophane film (pure regenerated cellulose) and his work provides an insight into the factors involved. The chemical and physical uniformity of the substrate together with its transparency greatly simplify the technical difficulties. Breakdown of cellophane buried in a number of soils was studied and although there were variations in detail from soil to soil, a similar general pattern of colonization emerged. The first colonizers were fungi and three groups, based on vegetative morphological characters, are noteworthy. The members of the first group were characterized by coarse mycelium that ramified over the surface of the cellophane and rapidly initiated decomposition by lysis of the cellulose adjacent to their hyphae. Such fungi, mostly in the form-genus *Rhizoctonia*, were clearly strongly cellulolytic and were dominant organisms on the substrate at this stage. In some soils, members of a second group, often species of the well-known cellulolytic genera *Humicola*, *Botryotrichum* and *Chaetomium* appeared as codominants. These fungi did not ramify extensively in contact with the cellophane but rather penetrated its thickness at scattered sites by means of 'rooting' hyphae which then branched out to form an approximately circular hyphal system within the thickness of the cellophane. Some lysis occurred at these sites, but disintegration was not extensive because the areas of penetration remained discrete even though the superficial mycelium was continuous. The third group of initial colonizers consisted of chytrids that commonly developed on some cellophane pieces.

The presence and activity of these initial colonizers were

D

obviously related to their ability to produce cellulases. Other factors, however, must have been involved because the soils contained many other cellulolytic fungi which rarely appeared on the cellophane. The extent of lysis produced by the various fungi obviously depended upon physiological and morphological factors. Chytrids were restricted by their rhizomycelium to very local activity although there lysis was intense. In theory, a high reproductive rate might counteract the small sphere of influence of the unicellular individual (as it does for bacteria) but reproduction in chytrids is comparatively slow and only the reproductive phase is motile. On the other hand, the power of linear extension conferred by hyphae, allied with a high growth rate and strong cellulolysis, made the *Rhizoctonia*-like fungi important agents of decomposition. The different morphology of the assimilative hyphae of the coarse mycelial and the 'rooting' groups influenced their activity but the basis of the difference is not known.

Bacteria were relatively uncommon during this initial fungal phase and appeared to cause little lysis of the cellophane. During mycelial senescence, however, they rapidly increased and presumably utilized either materials diffusing from the hyphae or the hyphae themselves. In turn, the bacteria supported a population of nematodes and protozoa, the former being parasitized by various predaceous fungi.

After the microorganisms had colonized the cellulose, mites, springtails and enchytraeid worms became active and the substrate became comminuted and unrecognizable by passage through their guts. Some cellulose was present in the excreta but its subsequent decomposition was not studied. Cellophane decomposition thus involved a wide range of microorganisms and small animals and occurred most thoroughly when both *Rhizoctonia* and enchytraeid worms were active.

Tribe's work demonstrates the complex pattern of colonization and utilization of even the simplest substrate. Any study which neglects a major component of such a natural system must be open to the criticism of unjustifiable simplification. Unfortunately, there are almost no comprehensive studies and the biochemical role of the soil fauna is largely unknown. It is, therefore, necessary for us to consider fungi *in vaccuo*, as it were, but it should be remembered that the picture presented will necessarily be incomplete.

From a brief review of nutrition presented earlier, it is clear that carbohydrate biochemistry is likely to be a key factor in ecology and that the diverse chemical substrates in soil (Gupta, 1967) will probably support an equally diverse microbial population. Precise experimental evidence of the role of biochemistry in fungal ecology is, however, hard to find although there is much circumstantial evidence.

If carbohydrate is so important, then competition for generally assimilable sugars and related products will be keen. The situation is particularly critical for the 'sugar' fungi, mainly members of the Oomycetes and Zygomycetes but with representatives in other classes. These fungi are characterized by an inability to utilize anything but the simplest carbohydrates, usually sugars but including starch. As all other fungi can assimilate these, too, the simplest hypothesis suggests that the 'sugar' fungi must occupy the substrate prior to all others if the simple carbohydrates are still to be there for their use. Prior occupation implies rapid germination of spores and either high growth rate or very abundant distribution and efficient dispersal.

Following the colonization of the complex substrate by sugar fungi, it might be supposed that there would be successive waves of colonization by fungi that utilize progressively more refractory components. Lignin decomposers, utilizing a substrate that relatively few organisms can degrade, would not be disadvantaged by a slow growth rate and less abundance for their nutrient source is persistent and extensive. Indeed, a slow growth rate might well be imposed upon them by the slow rate of enzymic actions breaking down the lignin to the basic monomers. Our hypothesis would thus lead us to expect a succession in time from sugar fungi to decomposers of cellulose or comparable polysaccharides and finally to decomposers of lignin, tannin, keratin and other substituted derivatives of carbohydrates. Biochemical analysis of decomposing leaf litter supports such a view because the order in which the components decrease is soluble sugars, hemicelluloloses, cellulose and finally lignin (Burges, 1958).

This biochemically based succession has been thought to be reflected in a taxonomic one, in so far as the Phycomycetes have been considered to be sugar fungi *par excellence*, the Ascomycetes and Fungi Imperfecti to be frequently cellulolytic, and lignin decomposition to be primarily the work of the higher Basidiomycetes (Garrett, 1951). More recent work, however, has shown that Fungi Imperfecti are important in the degradation of lignin, tannin and related compounds (Domsch, 1960; Jones and Farmer, 1967; Lewis and Starkey, 1969).

A classical succession apparently supporting these ideas is that on dung. If fresh dung is placed under a belljar to maintain high humidity, a succession of fungal fructifications can be observed. The first to appear are zygomycetes, followed by ascomycetes and basidiomycetes, and in general terms these have been considered to represent sugar, cellulose and lignin utilizers, respectively. Harper and Webster (1964), however, while confirming the sequence, have shown that it is not a succession based on nutritional factors. They grew a number of the fungi involved under a variety of conditions and showed that each had a characteristic minimum time between commencement of growth and the appearance of fruiting structures. If the fungi are listed in the order based on these minimum times, the sequence is found to be the same as that occurring on dung. The succession of fruiting bodies on dung is thus connected with the duration of necessary developmental periods rather than with differences in assimilatory ability.

A sequential study has been made of the colonization of sterile human hair placed on the surface of various soils (Griffin, 1960). In general, the first colonizers were *Fusarium* spp., certain *Penicillium* spp. and members of the Mucorales, and most of these are known to rely largely on simple sugars. These fungi were overlapped or followed by a second wave, of which *Chaetomium cochlioides*, *Humicola* spp., *Gliocladium roseum* and other *Penicillium* spp. were characteristic. Both *Chaetonium* and *Humicola* are genera with high cellulolytic ability and it seems probable that some of this second wave were utilizing polysaccharides. The final colonizers in all soils were keratinolytic members of the Gymnoascaceae or related imperfect genera. In the colonization of hair there is thus an indication of the anticipated succession based upon nutrition and it is most clearly shown in the regular late predominance of species utilizing the most resistant component of the substrate. An examination of the complete data, however, shows that the direct effect of the chemical nature of the substrate is certainly not the only one affecting the succession. English (1965) has suggested that the physical heterogeneity of hair may be of importance and that fungi producing boring and frond-like mycelium, and thus able to exploit the cortex at depth, might occur later in the succession than those limited entirely to the hair surface, a zone available to all fungi. The data on colonization give some support to this concept, but again it is no more than one of many factors.

Microorganisms active at one stage become potential substrates themselves and their parasites can only be distinguished with difficulty from other fungi attacking the original substrate simultaneously. Parasitic mucoraceous fungi frequently parasitize fungi, particularly other members of the Mucorales, growing on dung, and in many cases the relationship can be discerned. Utilization of previous members of a succession is, however, often difficult to demonstrate but it almost certainly occurred in the study on hair already noted.

There is another complicating nutritional factor of great and widespread significance. Cellulases and comparable lytic enzymes act externally to the cell and it is the sugar monomer that is absorbed and utilized. There is, therefore, a real sense in which organisms on cellulose, for instance, compete not for the cellulose but for the sugar monomers. It would thus seem theoretically possible for a non-cellulolytic fungus to grow on cellulose as long as it is spatially and temporally associated with a cellulolytic partner and is able to absorb at least some of the sugars released by the activity of the latter. Such a case has been described by Tribe (1966), who found that the non-cellulolytic *Pythium oligandrum* grew on cellulose so long as it was introduced along with cellulolytic *Botryotrichum piluliferum*.

A more complex, but more natural example is provided by Frankland's (1966, 1969) study of the fungal succession on decaying petioles of *Pteridium aquilinum*. At the time senescent petioles were laid on the soil surface, their soluble carbohydrate content was only 1·2% and the C:N ratio was 200. The standing petioles had been colonized by fungi such as *Aureobasidium pullulans*, *Epicoccum nigrum* and members of the Sphaeropsidales and these were joined by soil-borne members of the Sphaeropsidales and Tuberculariaceae when the petioles were laid on the soil. Most of these fungi are known to break down cellulose or lignin. The poverty of the substrate in simple sugars at this stage was reflected by a near absence of phycomycetes. In the second and subsequent years of decomposition, basidiomycete mycelium was common and the lignin decomposing *Mycena galopus* was especially prominent.

Hyphomycetes predominated in the third and fourth years but phycomycetes reached a marked maximum in the fifth and sixth years. During this late phycomycete phase the C:N ratio was reduced to about 50 but soluble carbohydrates were still less than 1%. The prominence of phycomycetes indicates, however, that simple sugars must have been available at this stage, even though they were assimilated as soon as they appeared. The source of the sugars is indicated by the results of subsidiary experiments in which *Mycena galopus* was grown in monoculture on petioles for one year. Initially, soluble carbohydrates made up 2·3% of the oven-dry weight and had decreased to zero by six months but there was an increase of 17% at one year. At the conclusion of the experiment, there were losses of 32%, 54% and 25% for α-cellulose, hemicelluloses and lignin, respectively. In the later stages of the natural succession it thus seems probable that the breakdown of complex carbohydrates by *Mycena galopus* and associated fungi gave rise to a small but constant liberation of simple sugars which were assimilated by mucoraceous fungi. A *Cephalosporium* abundant on bracken litter in the third year, failed to attack bracken in monoculture. It, too, may have been assimilating degradation products produced by the activity of other fungi. However, rapid colonization of mite pellets by this species suggests that it was mainly associated with the waste products of the meiofauna associated with decomposition.

### 2.2.2. *Zymogenous and autochthonous microorganisms*

The populations of fungi involved in the decomposition of ephemeral substrates fluctuate greatly and, following Winogradsky's ecological classification of soil bacteria, are said to be 'zymogenous'. In contrast, bacteria, and fungi which maintain a relatively uniform level of population and activity, regardless of the periodic incorporation of readily utilizable substrates, are said to be 'autochthonous'. The difference between the two categories is supposed to lie in nutritional factors, the zymogenous group being able to utilize only the simpler substrates whereas the autochthonous group is able to utilize more resistant substrates and, in nature, is forced to use these alone because of intense competition for sugars and simple polysaccharides.

All soil fungi for which we have much information are zymogenous and one could be forgiven for doubting if truly autochthonous forms exist. It is interesting to note that at least one soil bacteriologist (Clark, 1967) has severely criticized the validity of any attempt to divide soil bacteria into constant zymogenous and autochthonous groups. He has stated that the autochthonous bacteria are those for whom the environmental and nutritional conditions needed to evoke a zymogenous response are still unknown.

If autochthonous fungi exist, they presumably utilize lignin, humic acids and other intracticable substrates that persist for years in soil. Such fungi are likely to be slow growing and intolerant of competition on the normal nutrient media and so are likely to be isolated but rarely. The soil homobasidiomycetes share these attributes and perhaps they approximate an autochthonous fungus but, as with bacteria, it may only be our ignorance concerning these species that suggest such an appellation.

# 3 The significance of microbial interactions

## 3.1. Types of microbial interactions

### 3.1.1. *Commensal and mutualistic associations*

In natural habitats, microorganisms of one species usually grow in close association with those of many other species. Sometimes the association may be without discernible mutual influence and is then termed 'commensal', but such are probably rare. Mycorrhiza are well-known examples of mutualistic symbiosis where both partners appear to benefit from the association. In others, for instance the association between *Pythium oligandrum* and *Botryotrichum piluliferum* noted above, one member obviously benefits whereas the other appears unaffected. Similar instances must abound in the complex successions on natural substrates. Although such mutualistic or semi-mutualistic saprophytic associations must be common, their study has been overshadowed by that of antagonistic associations in which the interaction between organisms is harmful to at least one of them.

### 3.1.2. *Antagonistic interactions*

A simple example of the importance of antagonism is furnished by the difficulty of establishing a fungus in a natural soil in which it does not normally occur. Establishment is easy, however, if the soil is sterilized before the inoculum is added. Although sterilization usually produces significant changes in the nutrient status of soil, there is ample evidence that the predominating factor permitting colonization by the alien fungus is the elimination of the other microorganisms.

A more complex but more instructive example of the importance of antagonism concerns the ascomycete *Ophiobolus graminis*. This fungus causes take-all disease of wheat but also attacks many other monocotyledons, usually invading their roots. Although host infection readily occurs from hyphae lying in the infected remains of previous crops, infection by ascospores is rarely successful. Infection by ascospores has, however, been recorded, or is strongly indicated, in plants growing in sterile soil, in soil partially sterilized with methyl bromide, in cut-away peat, in recently deglaciated terrain and in newly drained Dutch polders (Gerlagh, 1968). Infection of the exposed, proximal part of seminal roots of wheat seedlings sown on the surface of natural soil and of roots lying in crevices between large soil crumbs has also been demonstrated (Brooks, 1965). The common factor is an environment of reduced antagonism, apart from which the inoculum potential of the ascospore is inadequate (Garrett, 1956a).

Take-all disease gradually decreases after a few years if the site is sown annually to wheat. This unexpected phenomenon of take-all decline appears to be attributable to increased antagonism to *Ophiobolus graminis* during both its saprophytic and parasitic phases. This specific antagonism appears to be induced by the continued presence of pathogenic strains of *O. graminis* themselves, for wheat growing, in the absence of the pathogen, is ineffective (Gerlagh, 1968). These examples of the influence of antagonism on the activity of *O. graminis* reveal little of the precise mechanisms involved and to this aspect we must now turn.

The literature on antagonism is large with much semantic confusion and considerable abstruseness. Indeed, one might associate this aspect of ecology with those definitions noted by Clark (1968): *ecology* is 'that branch of biology entirely abandoned to terminology' and *microbial ecology* is 'the art of talking about what nobody really knows about in a language that everyone pretends to understand'. Here, I shall adopt a somewhat simplified approach and present a viewpoint that appears to me to be indicated by the weight of evidence. For a more detailed knowledge, however, of the opposing viewpoints and the experimental evidence it is essential that the reader should refer to the many articles which may be traced through the reviews by Clark (1965) and Park (1960, 1967).

Antagonism may be attributed to exploitation, competition or antibiosis but *in vivo*, as we shall see, it is unlikely that any of these components act independently. Exploitation implies the utilization of the cells of one organism by another, the exploited cells being either living or dead, but if the latter, killed by the exploiter. Of many conflicting definitions, we shall here consider competition to be a rivalry for a factor that is in short supply in the environment and antibiosis to occur when the activity of one organism is restricted by an external metabolite of another. The meaning of the term 'antibiotic' has become increasingly restricted with time and is now applied almost exclusively in much scientific literature to complex, highly potent molecules often of high biological specificity. In the sense of our usage any external metabolite, whether it be as simple and general as carbon dioxide or as complex and restricted as griseofulvin, is a potential cause of antibiosis, and hence an antibiotic.

*Exploitation and lysis.* We have noted earlier that some soil animals, particularly mites and collembolans, feed largely on fungal spores and that many fungal structures are ingested by worms. Nematodes, too, pierce fungal walls and extract the protoplasm. In none of these cases of exploitation by animals is there any idea of their quantitative or qualitative effects on soil fungi: it is simply known that they occur (Boosalis and Mankau, 1965).

Fungi exploit other fungi, the interaction usually being termed parasitism (Barnett, 1964; Boosalis, 1964). Although biotrophic relationships are known for extraterrestrial mycoparasites, and for a few within soil (Barron and Fletcher, 1970; Dayal and Barron, 1970), a cruder necrotrophic relationship predominates among soil fungi, where the exploiter in one interaction may be the exploited in another. Mycoparasitic capabilities have been demonstrated for such common soil fungi as *Trichoderma viride*, *Rhizoctonia solani*, *Trichothecium roseum* and *Penicillium* spp. This might lead one to suppose that mycoparasitism is of common occurrence in soil. Most, although not all, workers have reported, however, that mycoparasitism by necrotrophic fungi is most intense under conditions of high nutrient status and at temperatures in excess of 25°C. Such temperatures are higher than those normally found in soil. Most of our knowledge has been obtained from *in vitro* experiments where even under optimal conditions mycoparasitism is rarely intense. In a field study, only six hyphae of *R. solani* out of 15,000 were found to be parasitized, and it is unlikely that further field evidence will reveal mycoparasitism to be a factor of great ecological importance.

There is increasing evidence that bacteria can penetrate the walls of fungi and multiply within them (Old, 1969; Old and

Robertson, 1969, 1970*a*, *b*) and it is likely that soil streptomycetes can do the same. Most commonly, however, cell wall lysis and disintegration of the cytoplasm occur while the bacterium or streptomycete is still outside the wall (Huber, Anderson and Finley, 1966). Here, however, it is difficult if not impossible to distinguish adequately between exploitation, competition and antibiosis (Lloyd and Lockwood, 1966). Ko and Lockwood (1970) have shown that the activity of β-D-glucosidase and chitinase of living mycelium of three fungi increased rapidly as the mycelia were leached by sterile, running water. The activity of these lytic enzymes was directly correlated with susceptibility of the hyphae to lysis. Several streptomycetes were able to lyse living, but not dead, hyphae. Mycolysis in the presence of streptomycetes was thus attributed to autolysis induced by deprivation of nutrients through the activity of the streptomycetes. There was no evidence for the action of antibiotics. Even in agar culture, half of the eighteen streptomycetes tested produced inhibition zones against fungi by nutrient depletion rather than by antibiotic production (Hsu and Lockwood, 1969).

It would be rash, however, to assume that all mycolysis is induced by energy depletion, even though this may yet be shown to be the major cause. Lysis, *in vitro*, can also be caused by oxygen depletion, toxic metabolites (self-produced or from other organisms) and by unfavourable *p*H (Brian, 1960). Some, at least, of these may be active *in vivo*. Potgeiter and Alexander (1966) have shown that strains of bacteria and actinomycetes are capable by themselves of inducing mycolysis *in vitro* producing such enzymes as β-1,3-glucanase and chitinase. Resistance to lysis in *Aspergillus nidulans* was directly correlated with the melanin content of the mycelium and resistance in other fungi appears to be often associated with melanization (Kuo and Alexander, 1967).

*Fungistasis, antibiosis and competition.* Examination of fungi in soil, whether by observations of sections or smears, by plating out or by respirometric studies, shows that most are inactive, existing mainly in the form of spores but sometimes as resting hyphae. This exogenous dormancy of much of the soil population has been termed 'fungistasis', or if by a classical purist, mycostasis. Opposing theories attribute fungistasis to competition and antibiosis and the subject therefore serves as a fitting general introduction to these two aspects of antagonism. The basic facts concerning fungistasis are as follows (Lockwood, 1964; Jackson, 1965). Fungistasis is widespread in natural soils, is non-specific, disappears in the absence of microorganisms, and is temporarily annulled by the addition of nutrients. There is general agreement that fungistasis is of microbiological origin because it is heat labile, destroyed by sterilization and reduced by factors that reduce biological activity such as low $pH$ and low temperature. Subsoils are rarely fungistatic. Furthermore, fungistasis can be restored to a sterile soil by mixing-in a little unsterile soil. Despite this basic agreement, there are many differences in detail between the results of experiments. These differences are perhaps attributable, in the main, to the use of a wide variety of assay techniques of dubious equivalence.

Non-sterile soil extracts and diffusates are often fungistatic and this was originally thought to imply the presence of water soluble inhibitory chemicals produced as the result of microbial metabolism. In short, fungistasis was attributed to antibiosis. Despite much effort, it has proved impossible to demonstrate the presence in soil of specific inhibitory external metabolites with the necessary properties. This failure led to the suggestion that fungistasis was akin to the staling phenomenon observed in fungi grown on nutrient agars. Staling is thought to result from the accumulation of a range of external metabolites that increase in concentration as the medium becomes largely colonized. Eventually, these metabolites become deleterious, inhibit germination and cause morphological changes in hyphae (Park, 1961b). It has been suggested that an analogous accumulation of metabolites might occur in soil because of the growth of microorganisms but confirmatory evidence has proved hard to find. With one exception (Balis and Kouyeas, 1968), evidence for any inhibitor is circumstantial and there are contra-indications. Thus leaching the soil with water or buffer solutions fails to reduce fungistasis and sterile soil extracts are usually, but not always (King and Coley-Smith, 1969b), without activity. Heating, distillation, fine filtration and extraction with solvents all greatly reduce or remove fungistasis. The case against antibiosis has been stated concisely by Lockwood (1964) – 'To base fungistasis on the existence of inhibitory materials in the soil mass would demand these minimum requirements: (a) demonstration of inhibitory activity in extracts made with mild reagents from a wide range of soils, and (b) demonstration of inhibitory activity by such extracts against a wide range of fungi. These requirements have not yet been met.'

The inhibitor hypothesis postulates the diffusion of metabolites from microorganisms to the fungal spore. The main alternative hypothesis postulates the diffusion of nutrients away from the spore to adjacent metabolizing cells, predominantly bacterial (Ko and Lockwood, 1967; Steiner and Lockwood, 1970). It thus explains fungistasis in terms of competition. The steep negative concentration gradient of nutrients around the spore is supposed to stop germination both of spores dependent on exogenous nutrients and of some spores able to develop on endogenous reserves in the absence of a steep concentration gradient.

Evaluation of the competition hypothesis requires evidence on the nutrient relations of germinating spores. Unfortunately, there are few data on germination available from experiments where exogenous nutrients have been rigorously excluded (Cochrane, 1966). Furthermore, percentage germination varies with environmental factors, age of spore and density of the spore suspension; the role of any added nutrient is various (Cochrane; Trinci and Whittaker, 1968; Griffin, 1970). It does appear, however, that few fungi habitually found in soil have spores that can germinate readily and abundantly without exogenous nutrients. An exception is the ascospore of *Neurospora crassa* which has been shown under exacting experimental conditions to germinate solely on endogenous substrates and this is of great significance for these spores do not exhibit fungistasis. The concentrations necessary to promote germination of other species are but poorly known. Doubtless, they will differ with species, but are probably less than 10 $\mu$g glucose ml$^{-1}$ (Trinci, 1969). To ask if the soil solution contains such amounts of assimilable carbohydrate is irrelevant, for the concentration will depend upon the water content of the soil and, more importantly, the amount will vary from site to site. The soluble carbohydrate contents of the soil solution, as a mean value, will be increased by localized pockets of high nutrient status where fungistasis has been annulled. It is not technically possible to determine concentrations solely at sites where fungistasis is operative. A further complication is provided, as shown by Steiner and Lockwood (1969), by the fact that germination depends on an integration of solute concentration with time rather than concentration alone. Thus the longer the time a spore needs for germination, the longer is the time over which carbohydrate concentration must exceed the critical value.

Integrating the various factors involved in support of the competition hypothesis of fungistasis, we arrive at the following picture. A body of soil will exhibit fungistasis if the amount of exogenous nutrient available to fungal spores falls below a certain critical level. An insufficiency of nutrient may rarely be attributable to an absolute deficiency in the soil at the given site but usually is the result of competition, whereby bacteria (with a far higher potential respiration rate than fungi) reduce the concentration of carbohydrate below the critical level within the germination period of the spore. Thus tardy germinators would be more susceptible to fungistasis, an exception borne out by experiments.

Bacteria on or near the spore surface may even cause such a negative concentration gradient that spores lose nutrients, thus accounting for fungistasis in fungi reported to germinate solely on endogenous nutrients in pure culture. Schuepp and Frei (1969) have shown that nearly all of the twenty-four species of fungi they tested were more affected by fungistasis as soil reaction changed from $pH$ 5 to $pH$ 7. On the nutrient hypothesis, such a result is not unexpected because of the marked decrease in bacterial respiration as $pH$ declines from neutrality (Stotzky and Rem, 1966, 1967; Stotzky, 1966*a*, *b*). Although current evidence gives considerable support to this nutritional or competition hypothesis, it would be rash to assume that nutrition is the only factor and that inhibitors have no role (Hawthorne and Tsao, 1969).

The germination of chlamydospores of *Fusarium solani* f. *phaseoli* and subsequent growth of the hyphae in soil presents an interesting example of the influences of nutrients and competition. In soil, chlamydospores required exogenous available carbon and nitrogen sources for germination and usually the former was the principal limiting factor. The amounts of exogenous nutrients needed to cause germination varied from soil to soil, and thus with the biotic environment, suggesting the importance of competition. In soils treated with antibiotics to reduce antagonism by bacteria, 30% germination of chlamydospores occurred in the absence of added nutrients (Cook and Schroth, 1965). In accord with these results, 60% germination occurred adjacent to seeds of *Phaseolus vulgaris* which exude many nutrients, including twenty-two amino acids, glucose, fructose and maltose during their germination.

Within 80 hr, however, nearly all the germ tubes had lysed. Bean hypocotyls exude sugars but only trace amounts of amino acids: adjacent to them, only 20% of chlamydospores germinated, but these produced persistent germ tubes and hyphae which formed thalli on the hypocotyls (Cook and Snyder, 1965). Clearly, amino acids stimulated germination but they also stimulated a microbial flora antagonistic to the germ tubes. Infection of the host was greatest adjacent to hypocotyls rather than seeds or root tips. The first organ alone exuded nutrients sufficient to induce germination but insufficient to stimulate antagonists to an effective degree.

The relative importance of antibiosis and competition, which we considered first in connection with lysis and then in regard to the generalized phenomenon of fungistasis, must be considered whenever antagonism occurs between microorganisms.

Antibiosis is best considered in two sections, related to the two concepts of an antibiotic. Despite much effort it has not proved possible to show the presence of specific antibiotics in the limited, popular sense in unaltered field soils. Such a failure is not unexpected for antibiotics are known to be readily inactivated or destroyed by a range of chemical and biological actions and by absorption on to the surfaces of clay particles. Furthermore, such compounds are usually produced in greatest concentration when there are ample nutrients, a condition not to be found in the bulk of the soil. Generally distributed specific antibiotics can be found in heavily amended soil, an example being provided by patulin produced by *Penicillium urticae* in stubble mulched soil (Norstadt and McCalla, 1969). Gliotoxin has been demonstrated in high concentration in straw and seed coats in an acid soil which naturally contained the producing organisms, *Gliocladium virens* (Wright, 1956). There is thus evidence for the occurrence of specific antibiotics in pieces of natural substrate that provide sites with a relatively high nutrient status. It is in such sites, where active colonization is occurring, that production of inhibitors might favour the producer in its competition with other organisms. The production of specific antibiotics in such localized sites is probably widespread but needs confirmation.

Antibiosis as considered above has involved more or less readily diffusable chemicals. Ikediugwu and Webster (1970a, b) and Ikediugwu, Dennis and Webster (1970) have demonstrated the existence of an interference between fungi that operates only when the hyphae of the two species are actually or nearly in contact. Hyphal interference involves loss of hydrostatic pressure, vacuolation and granulation of the cytoplasm and is presumably caused by a chemical scarcely able to diffuse through normal substrates.

It is facile to assume that production of a specific antibiotic will confer an advantage on the fungus producing it. This need not be so. Park (1960) has concisely stated another possibility which may well represent an approach to some field situations:

'$A$ inhibits $B$, but $C$ and $D$ are tolerant of $A$'s antibiotic. $B$, when active, inhibits $C$ and $D$, but has no adverse effect on $A$. In the presence of $A$, of course, $B$ neither grows nor produces its antibiotic, and so $C$ and $D$ grow. $A$, therefore, shares the substratum with two organisms, whereas were it not to produce the antibiotic it would share it with only one organism.' Park has from such a basis, argued that tolerance of inhibitors may well confer a greater advantage than does producing them.

When one turns from specific antibiotics to more general inhibitory products of metabolism, Park's argument for the importance of tolerance has great force. To quote one example (Barton, 1961), small cubes of wood obtained from dead oak twigs were autoclaved in a 2% glucose solution and placed in an unamended soil containing a natural population of *Pythium mamillatum*. About 50% of the cubes were colonized by this rapidly growing sugar fungus. If, however, the sterile blocks were placed on a soil with no native *Pythium* population for one or more days before burying in the original soil, no colonization by *P. mamillatum* occurred. Is the failure to be attributed to competition for nutrients by the microflora established during the initial period of soil contact or to antibiosis by that same population? An apparently conclusive answer was provided by placing the glucose impregnated blocks on *Pythium*-free soil for three days, then soaking the block for 24 hr in 2% glucose solution and finally placing the blocks on the *Pythium*-containing soil. Despite the restoration of the block to approximately its original content of glucose, no colonization by *Pythium* occurred. Antibiosis is thus indicated and is probably attributable to general inhibitors for there was no evidence indicating the presence of specific antibiotics. Such intolerance of antibiosis combined with its limitations as a sugar fungus have led Barton to categorize *P. mamillatum* as a typical pioneer sugar fungus limited to the colonization of virgin substances.

As a concluding illustration of the various types of biological interactions in soil, we shall consider the case of mycorrhizal short leaf pine (*Pinus echinata*) growing in soil infested with the pathogen *Phytophthora cinnamomi* (Zak, 1964; Marx, 1969*a*, *b*, 1970; Marx and Davey, 1969*a*, *b*; Marx and Bryan, 1969). It is now known that in many circumstances the mycorrhizal condition confers benefits upon both partners (Gerdemann, 1969; Harley, 1969; Smith, Muscatine and Lewis, 1969). The higher plant receives an enhanced supply of such nutrient ions as phosphate and potassium from the soil through the mediation of the fungus while the fungus depends largely upon the higher plant for a supply of assimilable carbohydrate. Such relationships are noteworthy but lie outside the scope of this book.

The mycorrhizal state, however, also appears to confer resistance to pathogens to which the non-mycorrhizal root is susceptible. Short roots of *Pinus echinata* are extremely susceptible to infection by the zoospores and mycelium of *Phytophthora cinnamomi* but mycorrhizas, in which the fungal partner may be *Laccaria laccata*, *Leucopaxillus cerealis* var. *piceina*, *Pisolithus tinctorius*, *Thelephora terrestris*, or *Suillus luteus*, are either resistant or of greatly reduced susceptibility. Resistance may be attributable to the physical barrier of the fungal mantle being interposed between root and pathogen but, at least in the case of *L. cerealis* var. *piceina*, antibiotic action is indicated. Both in culture and in mycorrhizas, this fungus produces diatretyne antibiotics that are strongly inhibitory towards *P. cinnamomi*. Both mycorrhizas in which this fungus is a component, and adjacent non-mycorrhizal short roots, are resistant to infection by *P. cinnamomi*. This suggests the action of such a diffusable inhibitor. The sequence of microbial interactions does not cease here, however, for the formation of zoosporangia and the subsequent release of zoospores of *P. cinnamomi* appears to depend upon the presence of soil bacteria. Thus, the total system involves a series of interactions, both mutualistic and antagonistic, between higher plants, fungi and bacteria.

# 4 Survival of fungi in soil

If a fungus is to maintain itself it must make continuous growth or intersperse periods of activity with dormancy. We shall consider survival under the aspects of competitive saprophytic colonization, saprophytic survival and dormancy and neglect the important aspect of parasitism as being outside the scope of this book. The survival of microbial plant pathogens in soil has received particular attention from Menzies (1963*b*) and Garrett (1970).

## 4.1. Competitive saprophytic colonization

The noun 'competition' was defined in an earlier chapter as 'rivalry for a factor in the environment which is in short supply', in accord with general usage in ecology. The adjective 'competitive' in the mycological literature, however, is used in a much wider sense so that competitive saprophytic colonization refers to saprophytic colonization of a substrate by a mixed population. There, a range of interactions are possible, more strictly referable to exploitation, competition *sensu stricto* and antibiosis. Competitive saprophytic colonization is obviously one of the two normal forms of colonization (the other being parasitic), but the major study of this and related concepts commenced only three decades ago and its initiation is associated particularly with Dr S. D. Garrett and his associates (Garrett, 1956*a*, 1963, 1970). It is instructive to treat the concepts in the light of their historical development.

One of the major groups of cereal diseases are those caused by fungi that attack the roots and lower stems and have a reservoir in soil. An important question that soon faced investigators was whether such fungi could increase their inoculum by saprophytic growth in the absence of the host. The range of possible substrates was large, but it was felt that, in the context of a wheat field, wheat stubble was an obvious choice. In experiments, the fungus was grown on a particulate nutrient medium (maize meal sand) and this inoculum was then mixed, in various proportions by weight, with soil. Sterile lengths of wheat straw were then buried in the soil/

inoculum mixtures and the percentage of straws colonized by the test fungus was determined after a suitable interval. Representative data are given in Table 4.1. and reveal that *Fusarium culmorum* and

Table 4.1. *Saprophytic colonization of wheat straw by four root-infecting fungi (after Butler, 1953)*

|  | Percentage straws colonized | | | | | | |
|---|---|---|---|---|---|---|---|
| Percentage of inoculum in soil/inoculum mixture | 100 | 98 | 90 | 50 | 10 | 2 | 0 |
| *Fusarium culmorum* | 90 | 87 | 80 | 67 | 65 | 55 | 22* |
| *Ophiobolus graminis* | 98 | 46 | 8 | 10 | 2 | 0 | 0 |
| *Helminthosporium sativum* | 100 | 33 | 46 | 10 | 8 | 2 | 0 |
| *Curvularia ramosa* | 100 | 96 | 96 | 96 | 85 | 88 | 0† |

* Colonized by naturally occurring *Fusarium culmorum* in soil microflora.

† *Curvularia ramosa* does not occur naturally in English soils.

*Curvularia ramosa* were far more successful in colonizing the substrate than were *Ophiobolus graminis* and *Helminthosporium sativum*. This fact was expressed in the statement that the first two species had a high competitive saprophytic ability whereas that of the last two was low. Competitive saprophytic ability was defined as 'the summation of physiological characteristics that make for success in competitive colonization of dead organic substrates' (Garrett, 1956a). Success in competitive colonization was then said to depend upon three factors: the competitive saprophytic ability of the fungus, its inoculum potential and environmental conditions, both animate and inanimate. The original concept of competitive saprophytic ability so derived has, in my opinion, been enormously useful and has been an integrating factor in much of the literature of soil ecology. Most scientific concepts, however, evolve in the light of new data from their simple beginnings and competitive saprophytic ability is no departure from this generalization.

It is to be noted that Garrett conceived competitive saprophytic ability as akin to pathogenicity in that both reflected an innate ability of the organism and presupposed in normal usage a particular substrate. When the competitive saprophytic ability of a pathogen was under consideration the implied substrate was logically that of the dead host. With the acceptance of the concept by other workers and its application to fungi which were obligately or primarily saprophytic, it soon became obvious that the substrate needed to be stated explicitly. To take an extreme case, *Nannizzia gypsea* and related fungi are successful in competitively colonizing only substrates composed primarily of keratin or other unusual proteins. By a direct extension of Garrett's original ideas, it can be said that their competitive saprophytic ability is high, for example, on hair. The specification of the substrate is essential for without it the statement would be quite misleading in that these fungi are unable to compete on the great majority of substrates. A very different example is provided by many sugar fungi whose competitive saprophytic ability we tend to think of as being high because of their rapid competitive colonization of nutrient agar media. Reflection, however, shows that this high ability is in the context of substrates with ample simple carbohydrates: their ability on heavily lignified material, for instance, is negligible.

As the chemical composition of the substrate influences colonization, so do other environmental factors such as presence of herbicides (Wilkinson and Lucas, 1969), temperature and water potential. The precise composition of the antagonistic microflora is also pertinent, for the presence or absence of certain antagonists may cause failure or success in colonization. The final position is thus: it is impossible to measure competitive saprophytic ability as an innate character in Garrett's original sense, because its measurement must always be under certain conditions of substrate and environment, which themselves profoundly affect the assessment. In all cases, the measure of competitive saprophytic ability is the measure of success in competitive colonization and this success clearly depends upon more than an innate characteristic of the organism.

Prior occupation of a substrate is a further factor which must be considered in colonization (Bruehl and Lai, 1966). If sterile hair is exposed on a soil whose flora contains no keratinolytic organisms, it is colonized by many fungi that utilize other chemical components of the hair. If, after a few months of such exposure, the hair is transferred to a soil with an indigenous keratinolytic organism such as *Nannizzia gypsea* then that fungus will colonize the hair, apparently unimpeded by the preceding organisms. A contrasting example is provided by *Fusarium oxysporum* and the colonization of sterile grass leaves or clover stolens. Park (1959) found that if these substrates were in contact with soil for three days before the opportunity for *F. oxysporum* to colonize occurred then that fungus failed to colonize. Its success at colonization is high only if it plays a pioneer role, with the opportunity to be present in the first wave of organisms.

In general, the ability of organisms to colonize substrates will vary with the amount of prior colonization and this can be a most important variable. In most, but not all, experiments the substrate has been sterile or at least very clean. In the field, however, such a condition is most unusual and potential substrates, such as leaves and stems, will be colonized by a variety of saprophytes and parasites even before they fall to the ground (Hudson, 1968). Such may be the basis of the discrepancy between some laboratory and field studies on the colonization of wheat stubble by *Fusarium culmorum*. Colonization often occurs in laboratory experiments where the straw is usually bright, clean and often sterile but appears to be rare in the field in the Pacific Northwest of the U.S.A. where the straw is weathered and already heavily colonized at the time it becomes a potential substrate for *F. culmorum* (Cook and Bruehl, 1968).

In previous chapters, we have considered some of the nutritional and antagonistic factors that influence colonization and we have just considered, in its historical perspective, the key idea of competitive saprophytic ability which it is clear must now be used with caution. It remains to provide a résumé of the factors involved in competitive colonization and their interactions.

A basic prerequisite is that the substrate shall either initially or by the action of other organisms provide the range of nutrients and particularly carbohydrates or other energy yielding compounds, required for growth by the specific fungus. The fungus must also be present on, or very close to, the substrate at the times when its nutritional requirements can potentially be satisfied. This, in turn, implies a certain minimal inoculum density. This required inoculum density need be very low if the fungus can utilize a compound assimilated by few others. If, however, the fungus is limited to the assimilation of compounds capable of being used by many organisms (simple sugars) it must colonize the substrate rapidly. This implies either a high growth rate, very high inoculum density or a combination of the two. Rapid germination of spores may also be a significant factor. It is possible that some fungi compensate for a moderate growth rate and inoculum density by the production of a wide range of enzymes that permit utilization of many carbohydrates of moderate complexity. On each substrate their colonization might be limited but in sum, over the many substrates in soil, large.

The chance of colonization by a fungus that produces relatively few spores may be enhanced by an ability to grow through soil from substrate to substrate. Fungi vary greatly in their ability to do this. It is a matter of common observation that basidiomycete mycelium ramifies through soil without continuous contact with discrete pieces of organic matter. *Verticillium alboatrum* grows readily through soil for a more limited distance from roots or infected debris (Sewell, 1959a; Heale and Isaac, 1963). On the contrary, hyphae of *Fusarium solani* and *Sclerotium cepivorum* make negligible growth away from their substrates (Scott, 1956a, b; Burke, 1965).

Although inoculum density has been emphasized above, in reality it is necessary to consider the energy status of the inoculum as well. A conidium which has lain dormant in the soil for many months and by endogenous respiration almost used up its reserves will obviously have a poorer chance of initiating growth than a newly formed conidium (Griffin and Pass, 1969). Similarly, a rhizomorph originating from a large substrate has been shown to be more successful in colonization than those originating from small ones (Garrett, 1956b) and hyphae growing from a propagule of *Rhizoctonia solani* smaller than 150 $\mu$m were unable to incite disease, no matter how many propagules were present (Henis and Ben-Yephet, 1970). It is the integration of inoculum density with a potential chemical energy factor that will normally be important and these factors are united in the concept of inoculum potential (Garrett, 1956a, 1960).

A puzzling point in connection with inoculum potential is whether the propagules should be conceived of as acting independently or synergistically (Garrett, 1960, 1966a). In those cases where unorganized mycelium is ineffective but hyphae aggregated into strands or rhizomorphs are effective, synergistic action seems to be indicated, but the meagre evidence available for spores indicates independent action. On a theory of independent action, increase in inoculum potential with increase in inoculum density implies that, in any instance, there are few sites on the substrate open to colonization by the fungus. Increased inoculum density then increases the chances of one spore being at such a site whereas other spores at inappropriate sites fail to contribute to colonization. Recently, mathematical models have been developed (Baker, Maurer and Maurer, 1967; Baker, 1968, 1970) that permit the prediction of the effects of inoculum density on colonization but very few data are available with which to test the validity of the models.

Other factors capable of affecting success in colonization are production of antibiotics (*sensu lato*), resistance to parasitism and tolerance of antibiosis: these have been discussed in earlier chapters. The relative significance of the various factors in influencing the outcome of attempted colonization of a substrate by a mixed population of organisms is uncertain. It is a scientific commonplace that the experimental technique used to investigate a problem determines to a considerable degree the answers obtained. This is so in the present case. In experiments on the colonization of fresh nutrient agar plates from mixed inocula, success was correlated primarily with rate of hyphal growth in pure culture (Wastie, 1961). Such was also the case when colonization of autoclaved soil from a mixed inoculum was assessed by noting the presence of fungi in sampling tubes, filled with potato dextrose agar, placed in the soil (Lindsay, 1965). When, however, the nutrient agar was partly depleted and staled by the prior growth of soil fungi, those fungi colonizing the agar from a newly added mixed inoculum were those most tolerant of the excreted products of prior metabolism (Dwivedi and Garrett, 1968). As noted earlier, the inability of *Pythium mammilatum* to colonize twigs in which there is an already established microflora is to be attributed to an intolerance of antibiosis rather than to a nutrient shortage. The effects of inoculum potential and of environmental factors have also been demonstrated by employing appropriate techniques (Lai and Bruehl, 1968).

Gibbs (1967) has studied the colonization of fresh pine logs and although these are atypical soil substrates, the work is considered here as an unusually clear demonstration of a number of the factors involved in saprophytic colonization and succession. The data are presented in Tables 4.2 and 4.3 and indicate that four factors, at least, are involved in the observed succession *Leptographium*

Table 4.2. *Factors associated with competitive saprophytic ability*
*(after Gibbs, 1967)*

| | | Rate of growth (mm day$^{-1}$) | | Ability to decompose | | Replaced in wood by |
|---|---|---|---|---|---|---|
| | 3% Malt Agar at 22·5°C | Fresh pine logs | Fresh pine stumps | Cellulose | Lignin | |
| *Leptographium lundbergii* | 10·0 | 4·0 | 8·5 | − | − | *F, P, T* |
| *Fomes annosus* (*F*) | 6·5 | 3·2 | 2·4 | + | + | *P, T* |
| *Peniophora gigantea* (*P*) | 6·8 | 1·8 | 2·2 | + | + | *T* |
| *Trichoderma viride* (*T*) | 10·5 | 0 | ... | + | − | ... |

Table 4.3. *Ability of fungi from the succession on pine logs to produce and tolerate antibiotics on 3% malt agar (after Gibbs, 1967)*

| Fungus previously grown on the agar | Linear growth (mm in 2 days) | | | |
|---|---|---|---|---|
| | *L. lundbergii* | *F. annosus* | *P. gigantea* | *T. viride* |
| None (control) | 16·0 | 9·9 | 10·2 | 21·0 |
| *Leptographium lundbergii* | 15·5 | 9·3 | 9·8 | 22·0 |
| *Fomes annosus* | 2·9 | 9·5 | 11·0 | 22·5 |
| *Peniophora gigantea* | 4·9 | 9·5 | 10·5 | 24·0 |
| *Trichoderma viride* | 0 | 0 | 1·3 | 14·5 |

*lundbergii* ⟶ *Fomes annosus* ⟶ *Peniophora gigantea* ⟶ *Trichoderma viride*. Table 4.2 shows that the succession reflects decreasing ability to tolerate toxic substances, present in the substrate itself but presumably declining with time and with increasing colonization. The initial member of the succession is also unable to decompose lignin or cellulose and is thus presumably a sugar fungus, assimilating the scarce simple carbohydrates by virtue of early colonization and fast growth. Table 4.3 records the ability of fungi in the succession to produce and tolerate antibiotics. The fungi were placed on pieces of cellophane overlying 3% malt agar and, after growth, were removed by lifting the cellophane. The agar was then reinoculated with the same or a different species, its rate of growth then depending both upon the antibiotics produced by the initial colonizer and upon the tolerance of the second colonizer to those antibiotics. The order of the succession can be seen to be the order of increasing tolerance to the antibiotics produced by other members of the succession.

In the light of the examples considered, it seems unreasonable to argue for the general supremacy of any one factor in influencing competitive saprophytic colonization, for the predominating factor is likely to vary with the precise circumstances of each case.

## 4.2. Dormancy

When a fungus is not actively colonizing a substrate, it may survive in the form of dormant structures such as spores, resting hyphae or sclerotia. To simplify a later argument, we shall assume that dormancy implies, *inter alia*, respiration based entirely upon endogenous reserves.

Sussman (1966) has distinguished two types of dormancy. In constitutional dormancy, development is prevented by innate characteristics of the organism. This may be associated with membrane impermeability, nuclear change or some other morphological or physiological factor. Examples of constitutional dormancy are afforded by oospores and zygospores and perhaps by those sclerotia where a weakening of the wall appears to be a prerequisite for germination (Chet, 1969). Exogenous dormancy is a condition imposed by an unfavourable environment and both chemical (e.g. nutrient deficiency) and physical (e.g. low temperature) factors may be involved. An essential characteristic of exogenous dormancy is that the process of germination commences immediately the propagule is removed to a favourable environment. The dormancy of most asexually produced spores and of resting hyphae is of the exogenous type and is intimately concerned with the phenomena of fungistasis discussed earlier.

Survival in the dormant state will cease if the physical or chemical environment should become lethal, for instance by the occurrence of exceptional cold or heat or the application of a biocide. In the absence of such a catastrophic change, survival of dormant propagules will depend upon their resistance to parasitism and to the toxic chemicals in the normal soil environment and upon the maintenance of the necessary endogenous respiration. The latter will depend not only upon the respiration rate but also on the rate of loss of nutrients to the exterior. The rates of endogenous respiration vary widely, the $Q_{O_2}$ ($\mu l$ oxygen $hr^{-1}$ $mg^{-1}$ dry wt.) being 75 for the sporidia of *Ustilago sphaerogena* but 0·3 for the ascospores of *Neurospora crassa*. Neglecting all other factors, it is to be anticipated that the ascospores will survive for a far longer period in the dormant condition than will the sporidia. Resistance to exploitation and antibiotics is often associated with melanization and thick cell walls, as exemplified particularly by sclerotia (Chet, Henis and Mitchell, 1967; Chet, 1969; Chet, Henis and Kislev, 1969; Jones, 1970).

Although it may be suspected that the difference in survival between the thin-walled conidia and the microsclerotia of *Verticillium* spp. (Green, 1969) is associated both with respiratory rate and resistance to parasitism, the relative importance of the various factors known to cause death of dormant structures is little understood. Survival of the conidia of *Cochliobolus sativus*, for instance, is reduced by such environmental factors as high soil moisture content and high temperature but there is strong evidence for the importance of antagonism. In some soils, but not others, most conidia were dead within fifty days (Old, 1967). Examination of conidia recovered from such soils revealed loss of cell contents and septa, rupture or erosion of walls and the appearance of well-defined holes of about 2·5 $\mu$m diam. in the conidial walls (Old, 1969; Old and Robertson, 1969, 1970a, b). All these changes were associated with the presence of bacteria within the conidia.

Longevity of dormant structures is thought of most meaningfully in terms of a population rather than an individual because decline in a population tends to be logarithmic, so that neither the longevity of an individual propagule nor the maximum longevity of any propagule is of much significance. The best comparative measure is the half-life period, that is the period during which the population declines to half the original (Yarwood and Sylvester, 1959). Such data are available for few soil fungi.

Collated longevity data refer mainly to the longest period at which viable propagules could be found (Sussman, 1968). This is not an objective measure because detection of viability depends upon the sensitivity of the technique. Frequently, a host plant is the most sensitive detector of propagules of a pathogen. Such a situation is potentially serious in a perennial crop, such as bananas, where disease may become apparent in the second or later years after planting, initiated by a pathogenic population too small to be detected at the time of planting by the means available (Rishbeth, 1955).

Longevities recorded vary from a few days for some thin walled conidia and zoosporangia, to many years for thick walled chlamydospores, conidia, sclerotia and oospores. Generalizations are dangerous, however, for the apparently similar structures of closely related fungi can differ greatly in their survival. Thus few chlamydospores of *Fusarium solani* f. *cucurbitae* survived in a Californian soil for more than one year whereas survival of chlamydospores of *F. solani* f. *phaseoli* was scarcely diminished at 600 days (Nash and Alexander, 1965). Structures can also become modified in ways which influence their survival. Examples are provided by the progressive thickening of the walls of sporangia of *Pythium ultimum* in soil until they approach the thickness of oospore walls, the progressive melanization of the endoconidia of *Thielaviopsis basicola*, the fusion into 'aerial microsclerotia' of adjacent conidia of *Verticillium dahliae*, and the replacement of macroconidia by chlamydospores in *Fusarium solani*. This latter change has been shown to be induced by at least three chemicals in the neutral and anionic fractions of soil, the effectiveness of each varying among the different clones of *F. solani* f. *phaseoli*. The active chemicals appear to be bacterial products (Ford, Gold and Snyder, 1970*a*, *b*). Such early changes in morphology and perhaps physiology are probably of great significance. The initial decline in most populations is rapid. Apart from the inherent variability of a population, early loss may be caused by the death of those propagules in which the protective changes are tardy or prevented by particularly inimical conditions. That something of this sort is involved is suggested, for example, by the rapid decline in population of conidia of a soil streptomycete for which the half-life period was only twelve days. Conidia are, however, the main means of survival in streptomycetes and are, as a population, very persistent (Lloyd, 1969).

A final factor of note in connection with the efficiency of dormancy as a survival mechanism is that germination should not occur in the absence of a substrate (Garrett, 1970). Efficiency would thus seem to require that the activator be associated, preferably invariably, with the presence of a suitable substrate. Such an activator/substrate relationship is clear where sugars act as activators for the exogenously dormant spores of sugar fungi. In few cases, however, is the factor causing germination in natural soil known with any precision. The ecological significance of the prevalence of temperature extremes or fluctuations as activators is largely unknown (Warcup and Baker, 1963). Two cases in which the nature and mode of action of the activators have been studied extensively indicate the range of mechanisms involved. Germination of the chlamydospores of *Fusarium solani* f. *phaseoli* in soil (see p. 50) depends upon exogenous sources of simple carbohydrates and nitrogen sources and is thus very sensitive to antagonism. Antagonism is also involved as a factor determining germination in *Sclerotium cepivorum* but here the mechanism is different to that in *F. solani*.

*Sclerotium cepivorum* causes white root disease of onion. Its dormant stage is a sclerotium which rarely germinates in field soil except in the presence of *Allium* spp. Spontaneous germination does occur, however, under sterile conditions, whether on agar media or in soil. The difference does not appear to be a matter of nutrient availability for good, though variable, germination occurs in sterile situations where there are no apparent exogenous nutrients (Coley-Smith, King, Dickinson and Holt, 1967). The indications are that S-allyl-L-cysteine sulphoxide (alliin), S-*n*-propyl-L-cysteine sulphoxide (dihydroalliin) and related non-volatile, water soluble chemicals characteristic of *Allium* spp. diffuse from roots into the soil. There they are rapidly degraded by alliinase and related enzymes, produced by soil bacteria, to yield a range of volatile, water-insoluble alkyl sulphides, principally propyl and allyl mercaptans, sulphides and disulphides. These volatile sulphides show strong activity in stimulating sclerotial germination in unsterile soil (Coley-Smith and King, 1969; King and Coley-Smith, 1969*a*). Stimulation by extracts of *Allium* spp. occurs at concentrations too low to inhibit bacterial growth so that the effect is not likely to result from inhibition of antagonists. Keyworth and Milne (1969) on the contrary, have shown that onion exudates somehow induce tolerance in *S. cepivorum* to a range of antibiotics present in agar media. Thus the current picture is of a dormancy imposed by the inhibitory effects of microorganisms and broken by the breakdown products of host exudates acting to induce tolerance. Volatile chemicals from the initial stages of decomposition of plant remains also appear to act as stimulators for the germination of sclerotia of *Sclerotium rolfsii* (Linderman and Gilbert, 1969). Widely distributed compounds such as methanol, acetaldehyde and isobutyraldehyde are effective and this is perhaps appropriate for a fungus able to colonize many different plant species.

## 4.3. Saprophytic survival

Lying between competitive saprophytic colonization and dormancy is a shadowy state termed saprophytic survival (Garrett, 1970). It is exemplified particularly by a parasite surviving in tissues that were colonized when the host was alive. The state is to be distinguished from dormancy by the existence, albeit at a low level, of exogenous respiration and from saprophytic colonization by the failure to colonize new independent areas of tissue. None the less, new cells within the tissue previously colonized in general may well be colonized. Experimentally, these limits of the state cannot be established rigorously, so that the concept must remain nebulous though of practical importance.

Saprophytic survival has been studied most extensively in parasites of the roots and bolls of trees and in cereal parasites (Garrett, 1970). Survival in massive tree remains is fifty to one hundred years for *Poria wentii* (Childs, 1963) and decades for *Armillariella elegans* but is only a matter of months or a few years for fungi such as *Ophiobolus graminis, Cephalosporium gramineum* and *Gibberella zeae* in cereal tissues (Garrett, 1956a; Lai and Bruehl, 1966; Burgess and Griffin, 1968a). In trees, the fungus frequently develops a pseudosclerotium delimited by zone-lines of melanized bladder-like cells buried in the tissue of the substrate (Campbell, 1933, 1934; Rishbeth, 1951). Little, however, is known of the persistence and activity of individual hyphal cells within the pseudosclerotium. Garrett (1940) has observed microscopically an increase in the dark mycelium typical of *O. graminis* during its survival in wheat straw, but survival by hyphal dormancy also occurs there (Scott, 1969b).

Garrett has concluded that the economic utilization of the substrate is the key factor in prolonged saprophytic survival. Nyvall and Kommedahl (1970) agree that conditions favouring either the growth of microorganisms in general or the excessive growth of a particular species will reduce survival of that species. Garrett's conclusion is based largely on studies of the influence of available nitrogen on the persistence in straw of cereal parasites such as *Ophiobolus graminis* and *Cochliobolus* (*Helminthosporium*) *sativus*. These studies have shown that survival of *O. graminis* is enhanced by added nitrogen whereas the converse is true for *C. sativus*. Garrett (1967) has developed the following explanatory argument.

With a fungus of somewhat low cellulolytic ability, like *Ophiobolus graminis*, the original mycelium in the infected cereal straw will die fairly soon in the absence of extra nitrogen, because the hyphae exhaust their zones of enzymic erosion of the cellulose constituent of the host cell walls, and so perish from carbohydrate starvation. But if a supply of soluble nitrogen continuously diffuses into the infected tissue from the surrounding soil, then new branch hyphae are formed and these erode fresh areas of host cell wall, so that the life of the fungal colony is thereby prolonged.

'With a fungus of relatively high cellulolytic ability like *H. sativum*, on the other hand, the zones of enzymic erosion around individual hyphae are wider so the original mycelium can survive for longer on the soluble products (cellobiose and eventually glucose) of its cellulolytic activity. It has a lower need, therefore, for soluble nitrogen from the soil to permit the production of new branch hyphae. On the contrary, too high a level of nitrogen supply will cause the mycelium to branch too vigorously, with the result that the colony will decompose cellulose at a rate in excess of that required for survival; the result of this will be premature death of the fungal colony through early exhaustion of cellulose reserves.

'One further point needs elaboration. The effect of nitrogen on longevity of survival does not seem to be related simply and directly to cellulolytic ability alone but rather to cellulolytic ability of a particular fungus in relation to its need of soluble carbohydrate for respiration and growth.'

This argument, since elaborated further (Garrett, 1970), is most valuable in visualizing some factors of great potential significance. I find existing data on the influence of nitrogen to be suggestive but not conclusive, for linear growth rate has been the main measure of 'respiration and growth' (Garrett, 1966*b*). This I consider unfortunate because linear growth neglects the aspects of both respiratory efficiency and mycelial density which are likely to be significant and which vary with species and environmental conditions (Trinci, 1969). Furthermore, the argument does not consider the significance of possible differences in efficiency of the nitrogen metabolism of different species, as discussed above on pp. 33-4.

That nutritional factors, and thus competition rather than anti-biosis or exploitation, predominate in the saprophytic survival of *Ophiobolus graminis* has been shown by Scott (1969*a*). His data indicate that in most experiments, death was attributable to failure to utilize the straw substrate because of a deficiency of available nitrogen. Too much nitrogen, however, caused death through substrate exhaustion (straw breakdown) because of the enhanced activity of, presumably, both *O. graminis* and the soil microflora. The insignificance of antibiosis is indicated by the long survival of *O. graminis* in straws buried in soil amended with glucose and calcium nitrate in proportions to give a C:N ratio of 10. Such an amendment would surely have increased microbial activity and thus potential antibiosis yet there was no deleterious effect on *O. graminis* because the C:N ratio of the amendment was sufficiently low to increase available nitrogen even after the utilization of all the added glucose.

Garrett has considered antibiotic production to be of little or no importance in saprophytic survival in the fungi he has studied. Lai and Bruehl (1966), however, consider that *Cephalosporium gramineum* survives in moist straw pieces by maintaining a low rate of metabolism accompanied by production of an antifungal antibiotic. They attribute death of *C. gramineum* to its loss of dominance in the substrate when the concentration of available nutrients falls below the level necessary for effective antibiotic production. Their data do not, however, rule out the simpler 'starvation' hypothesis. Were it possible, it would be most instructive to compare sapro-phytic survival of two strains of *C. gramineum*, one producing and one not producing the antibiotic.

As stated at the beginning of this chapter, long-term survival in soil will be determined by a combination of the constituent factors discussed above. The relative importance of the three factors is quite unknown for most species but certainly differs not only from species to species but also from strain to strain and from one soil environment to another. Appearances can also be deceptive, for neither ability to grow and colonize organic fragments in soil nor the production of sclerotia were clear indicators of ability of strains of *Rhizoctonia solani* to survive in soil (Baker, Flentje, Olsen and Stretton, 1967).

# PART TWO  Physical Ecology of Soil Fungi

# 5 The influence of water

In the previous Part of this book, aspects of the ecology of soil fungi have been discussed which are of a basically 'chemical' nature. It has been chemicals, whether by presence or absence, which have stimulated or inhibited. The present Part considers the role of physical factors in the ecology of soil microorganisms, particularly of soil fungi. This is not to say that chemical and physical factors are clearly separable: they are not. Rather it is to say that in this Part systems will be considered where the physical components are at least as important as the chemical in determining biological activity. Because the physical concepts are relatively unfamiliar to many biologists and because such aspects of fungal ecology are not prominent except in the most specialized reviews (Griffin, 1963a, 1969; McLaren and Skujins, 1967; Cook and Papendick, 1970b), the treatment will be more detailed in this than in the previous Part.

As with the chemical factors in nutrition where, for instance, the amount of carbohydrate must be considered in relation to the amount of available nitrogen, so the various physical factors cannot be thought of in isolation. Bateman (1963) has provided an instructive analysis of environment and pathogens in relation to a root rot complex and others have noted interactions between temperature, $p$H and soil moisture in determining fungal activity (Griffiths and Siddiqi, 1961; Ward and Henry, 1961; Persson-Hüppel, 1963). At the present time, however, it is rarely possible to follow the effects of chains of factors or their interactions with any accuracy. Thus, it is necessary in experiments to simplify the system by attempting to reduce the number of variables. Hypotheses derived from such simplified systems may give false indications if they are applied incautiously to field situations.

The physical factors of most apparent importance to soil organisms are temperature, hydrogen ion concentration and a vast complex associated with texture, structure, water and gaseous exchange (Fig. 5.1). Before discussing them, however, it is

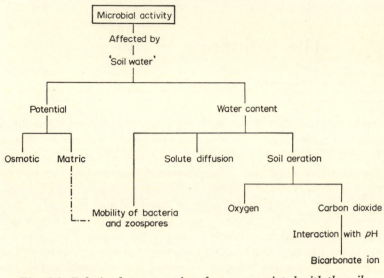

Fig. 5.1. *Relation between various factors, associated with the soil water regime, that affect the activity of microorganisms*

necessary to consider the various ways by which the effect of an environmental change on fungal growth can be assessed.

## 5.1. The measurement of fungal growth

The most appropriate measurement for expressing the effect of environmental variables on the growth of fungi has for long been a bone of contention. Some have argued that the rate of increase in total dry weight of the colony is most appropriate, others that the rate of linear extension of the colony margin is best. The correct answer is probably determined by the reason for which the data is required. Thus rates of synthesis of protoplasm or of respiration are likely to be correlated with increases in dry weight whereas rate of spread over the surface of a new substrate would be correlated with rate of linear extension. The most definitive study of the problem is that of Trinci (1969), who studied the relationships between the radial growth rates of colonies on agar media and the specific growth rates in submerged culture. (The specific growth rate, $\alpha$, is obtained from $dW/dt = \alpha W$ or, more conveniently, from $\alpha = \ln 2/t_d$, where $W$ is the dry mass of culture at time $t$ and $t_d$ is the doubling or 'mean effective generation' time.) The relationship between these two growth parameters differed greatly from one fungus to another (Table 5.1) and indicated that both *Penicillium chrysogenum* and *Aspergillus nidulans* synthesized new protoplasm quicker than *Mucor hiemalis* although their radial growth rates were far slower. None the less, Trinci concluded that, for a given fungus, the rate of radial growth of a colony is a reliable parameter to determine the optimum temperature for growth. Changes in the specific growth rate brought about

Table 5.1. *Relationships between different parameters for measuring growth of fungi (after Trinci, 1969)*

| Fungus | Colony radial growth rate ($K$, $\mu m\ hr^{-1}$) | Specific growth rate ($\alpha$, $hr^{-1}$) | $K/\alpha$ |
|---|---|---|---|
| *Aspergillus nidulans* | 146 | 0·148 | 986 |
| *Penicillium chrysogenum* | 53 | 0·123 | 432 |
| *Mucor hiemalis* | 424 | 0·099 | 4,296 |

by temperature changes were usually reflected in the radial growth rate.

Most of the data on the effects of environmental variables on the growth of fungi discussed in this book are based on the parameter of the radial growth rate of the colony, or, more generally, on the rate of linear extension. The use of this parameter is thought to be valid but it must be remembered that it neglects the aspect of mycelial density and so may be misleading if used in connection with total respiration.

## 5.2. Particles and voids

The individual solid particles of a soil vary greatly in size but they are predominantly microscopic and are more or less closely packed together. In between the solid particles is a complex anastomosing series of voids, or pores, filled with either an aqueous solution or gas. The sizes of these particles and voids have a profound effect on the water regime in the soil and hence on the soil atmosphere. Their sizes also have more direct effects and it is appropriate to consider these first. Thus, in a series of artificial soils composed of aluminium oxide particles of relatively uniform size, the intensity of sporulation of *Curvularia* spp. (length of conidium $-27\ \mu m$)

was reduced in voids of less than 46 $\mu$m diam, where scarcely more than one or two conidia could have been formed from each conidiophore (Griffin, 1963b). Similarly, production of oogonia (20 $\mu$m diam) of *Pythium ultimum* was abundant in pores of 45 $\mu$m diam but sparse in pores of 15 $\mu$m diam (Griffin, 1963c). *Trichoderma viride* sporulated less among glass beads where the voids were about 9 $\mu$m diam than where the voids were 39 $\mu$m diam (Parr and Norman, 1964; Parr, Parkinson and Norman, 1967; Griffin, 1969). All these data were obtained from systems in which there was an artificial uniformity of pore size and it may be that such spatial limitations on sporulation are unimportant in a natural heterogeneous soil. Kubiena (1938) noted the production of depauperate fructifications in soil pores but on the evidence of Sewell (1959a), it is possible that their size was affected by nutritional rather than spatial factors.

The size of the soil particles is of significance through its influence on the total internal surface area of the soil. There is now ample evidence that many chemicals, including nutrients and enzymes, become absorbed on to, or have increased concentrations adjacent to, solid surfaces in contact with dilute solutions. Such an effect is shown not only by complex solids, such as clays, but also by glass and other simple solids. Furthermore, bacteria attach themselves to surfaces (Zvyagintsev, Pertsovskaya, Duda and Nikitin, 1969), partially as a consequence of increased nutrient concentration but also because some appear to be inherently sessile. An increase in internal surface area consequent upon reduction in particle size is thus likely to stimulate bacterial activity and such effects have been recorded and analysed (ZoBell, 1943; Bhaumik and Clark, 1947; Estermann and McLaren, 1959; Parr and Norman, 1964; Parr, Parkinson and Norman, 1967). Fungal respiration, on the contrary, is greatest in systems with large particles. Reduced activity of fungi in small particle systems may possibly be attributed in some cases to poor aeration (Parr and Norman, 1964; Griffin 1963b, c, 1969) but not in others (Bhaumik and Clark, 1947).

Soil particles are usually aggregated into structural units, or peds. Between the peds lie large voids whereas within them, the bulk density is usually high because of the relatively close packing of the individual particles. Breakdown of these aggregates during tillage stimulates microbial activity, probably because of the exposure to microbial attack of hitherto protected surfaces (Rovira and Greacen, 1957; Greenwood, 1968a; Allison, 1968). Aggregation of the particles is in part the result of chemical cementing but also of microbial actions. The precise nature of this action is uncertain (Griffiths, 1965; Allison, 1968) but fungi, as well as bacteria, are implicated. Thus, Dommergues (1962) found that structural regeneration commenced at water potentials so low that bacteria, but not fungi, would be inactive.

The presence of microorganisms on soil particles sometimes leads to the production of organic films, which so alter the advancing contact angle of water that the soils in extreme cases become water repellent (Emerson and Bond, 1963; Bond, 1964; Bond and Harris, 1964; Debano, 1969). Basidiomycete hyphae, often hydrophobic themselves, were particularly associated with many such sites. It is likely that barren zones within fairy and other rings in grasslands are to be similarly attributed to the hydrophobic surfaces of hyphae and adjacent particles, leading to the death of higher plants from drought (Shantz and Piemeisel, 1917; Toohey, Nelson and Krotkov, 1965; Griffin, 1969).

The opacity of soil makes the direct observation of fungi within it difficult. Glass microbeads, however, provide a translucent system in which to study the effects of size of particles and pores (Parr, Parkinson and Norman, 1963). In such a system, partly filled with water, conidial production by *Curvularia* spp. and *Cochliobolus sativus* was limited to air-filled pores (Plate 1) whereas pionnotal masses of macroconidia of *Fusarium culmorum* were formed principally just beneath the air–liquid menisci (Griffin, 1968b). The significance of soil water is rarely so simple, however, and detailed consideration must now be given to the properties and microbiological importance of the aqueous solution held within the matrix provided by the solid particles.

## 5.3. Concepts
Many of the concepts, definitions and units applicable to the soil water regime have been precisely expressed in physical terms by the International Society of Soil Science. They are most easily found in the book *Agricultural Physics* (Rose, 1966). This book also contains brief descriptions of many relevant techniques and, indeed, should be taken as an unexpressed reference for the sections that follow.

### 5.3.1. *Water content*

The water held in a soil can be measured gravimetrically (as grams of water per gram of soil dried for 24 hr at 105°C, or g g$^{-1}$) or volumetrically (cubic centimetres of water per cubic centimetre of soil, or cm$^3$ cm$^{-3}$). The former measure is the more easily obtained but is the less satisfactory because of the shrinkage and expansion that accompanies change in water content in many soils. Such volume changes are accurately reflected in the volumetric, but usually are not in the gravimetric, measure. Change in the overall volume of a soil sample is best expressed in terms of bulk density, where bulk density is equal to the mass of solids in the sample divided by the total volume of all constituents – solids, liquids and gases. Gravimetric water content can be converted to volumetric by multiplying the former by the bulk density.

When all their voids are water-filled, different soils may contain very different quantities of water so that two soils each containing 0·2 g g$^{-1}$ water, may well differ in many important ways associated with their water regimes. In an attempt to reduce this difficulty, it has become conventional in much of the literature of soil mycology to express water content on a relative basis, that is as a percentage of the value when the soil is saturated (Griffin, 1966b). Thus, in experiments, it has thus been the practice to adjust samples of different soils so that each is, say, 60% saturated rather than to adjust each to the same absolute water content. Unfortunately, as we shall see, this adds little to the precision of the experiment.

Although absolute water content is an easily understood property of a soil, it is of little direct importance, as a property in its own right, so far as microorganisms are concerned. The volume of water needed for the creation of a few centimetres of hyphae or a few million bacteria is so small that it will be contained within one gram of many apparently dry soils. The microbiological importance of water content lies in the number of other properties with which it is correlated.

### 5.3.2. *Matric potential*

If a capillary tube is placed vertically with its tip dipping into water, water rises within it. At equilibrium, the forces associated with the interfaces between the water, the air and the glass walls

Plate 1. a. Curvularia *sp. growing among glass microbeads. Hyphal growth has occurred in voids filled with air and in voids filled with dilute nutrient solution; sporulation has occurred only in the former.*

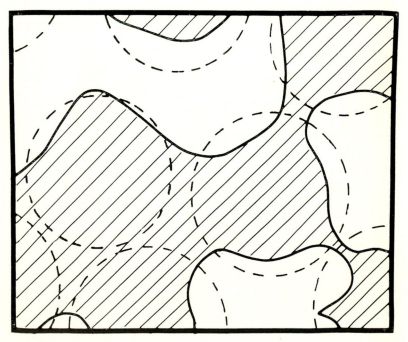

b. *Diagram of the system shown in* a, *with outlines of glass microbeads and the menisci separating air-filled from liquid-filled portions of voids (liquid denoted by shading).*

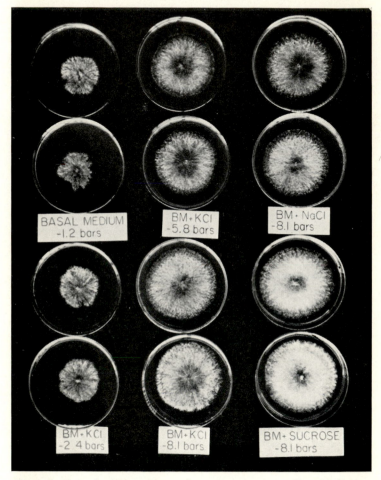

Plate 2. *Growth of* Fusarium culmorum *on agar media of different osmotic potentials.*

of the tube (surface tension) balance the downward force exerted by the column of water raised within the tube. More precisely:

$$-\pi r^2 h \rho g = 2\pi r \sigma \tag{1}$$

where $h$ is the height of water column, above the reservoir, within the capillary tube (cm); $\rho$, the density of water (g cm$^{-3}$); $g$, the gravitational constant (cm sec$^{-2}$); $r$, the radius of tube (cm); $\sigma$, the surface tension (dynes cm$^{-1}$).

If an organism external to the system is to absorb water from the air/water interface within the tube, it must overcome the forces tending to retain the water within the tube. In pressure terms, from Equation (1), this is equal to $-h\rho g$ dyne cm$^{-2}$. It can be said, therefore, either that the pressure of the water at the interface within the tube ($\tau$) is $-h\rho g$ dyne cm$^{-2}$ or, defining suction as negative pressure and changing units, that the water exerts a suction equivalent to a column of $h$ cm water.

As potential energy is a more fundamental concept than pressure, it is becoming usual in soil science to refer to the energy status of the soil water rather than to its pressure. The units of potential energy are erg g$^{-1}$ or joule kgm$^{-1}$ but when water, with a density of 1 g cm$^{-3}$, is being considered, potential can be expressed as erg cm$^{-3}$. These latter units are dimensionally the same as dyne cm$^{-2}$, the units of pressure. In the special case of water, therefore, the pressure and potential will have the same numerical magnitude and sign.

Following this argument, pressure and potential can be expressed in a variety of units, of which the bar is possibly the most convenient for the purposes of soil microbiology. The relationships between the commoner units are given by 1 bar = 10$^6$ dyne cm$^{-2}$ = 100 joules kgm$^{-1}$ = 0·987 atm = 1,022 cm water = 75 cm mercury. Suction is also sometimes expressed in logarithmic form, where $pFx = 10^x$ cm water. The relative humidity (R.H.) of the vapour phase in equilibrium with a body of water is also an indirect measure of the potential of the water. In the absence of solutes, the matric potential ($\tau$ bar) is given by:

$$\tau = (RT\rho \ln p/p_o)/10^6 M$$
$$= 10·65\ T \log p/p_o \tag{2}$$

Also:

$$h(\text{cm}) = (RT \ln p/p_o)/Mg$$
$$= 10,844\ T \log p/p_o \tag{3}$$

where $R$ is the gas constant; $T$, the temperature (K); $\rho$, the density of water at $T°$; $M$, the molecular weight of water; $p$, the vapour pressure of water under the given experimental conditions; and $p_o$, the vapour pressure of a reference pool of pure water at $T°$ ($p/p_o$ is thus the water activity often expressed as $a_w$, or the relative humidity expressed with base of 1, not 100). As a rough rule a decrease of 1% R.H. in the range 75 to 100% R.H. is equivalent to a decrease of 14 or 15 bar potential. The potential that we have been considering, although a property of the water, depends upon the radius of the capillary tube, as shown by Equation (1), which may be simplified to:

$$\tau \ (\text{dyne cm}^{-2}) = -h\rho g = 2\sigma/r \qquad (4)$$

More generally, the potential of a volume of water will depend on the geometry of the matrix within which it lies. In soil, the matrix is extremely complex and gives rise to a number of forces associated with water/solid or water/air interfaces (Day, Bolt and Anderson, 1967). All are physical phenomena depending upon the solid phase, or matrix, and they are, therefore, combined under the term 'matric potential'. The potential of the water derived from its presence within a capillary tube is a simple example of matric potential.

Consider now the soil sample depicted in Fig. 5.2*a* in which a soil, originally saturated, has lost some water by evaporation. The water is held in only a part of the total system of voids and in all cases, the air/water interfaces occur at situations where the radius of the pore is $1\cdot5 \times 10^{-3}$ cm. (In fact, the radius of curvature of the air/water interface in a soil is by no means simple because of the complex geometry of the boundary walls imposed by the matrix. When applying Equation (4) to a soil, $r$ is the 'effective' or 'equivalent' radius of a soil pore having the same effect as a capillary of radius $r$.) Further, from Equation (4) and knowing that $\sigma = 73$ dyne cm$^{-1}$:

$$h = -0\cdot15/r \qquad (5)$$

Therefore, it can be calculated that the soil water in the system depicted in Fig. 5.2*a* has a matric potential of $-100$ cm water. If further evaporation occurs, the system will become, say, as depicted in Fig. 5.2*b*. Once again, all menisci lie in pores of equal

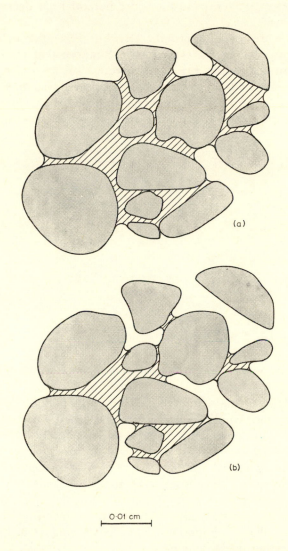

Fig. 5.2. *Section through soil sample, showing distribution of water at two matric potentials,* (a) −100 *cm water, and* (b) −200 *cm water*

radius ($7 \cdot 5 \times 10^{-4}$ cm) and the water potential has decreased to $-200$ cm water.

Soil pores are not of uniform diameter but tend rather to be vase-shaped with abundant cross-linkages. Suppose that in the simplified pore depicted in Fig. 5.3 (where $r_1 > r_4 = r_5 > r_2 > r_3$), water is draining from an originally saturated system. Suppose further, that when a matric potential is imposed, the meniscus is at the site where $r = r_1$. If the potential is decreased by evaporation

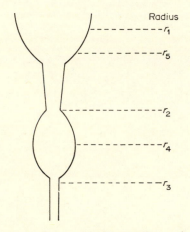

Fig. 5.3. *Section through an idealized soil pore* (*see text, p.* 80)

or by applying gentle suction to the base of the pore, the meniscus will retreat until equilibrium is again established – now in a pore where $r = r_2$. If a further decrease in potential occurs, the meniscus must now move through the entire pore, where $r = r_2$, until it reaches a small pore where $r = r_3$. Water, therefore, tends to leave the system in 'gushes' each time the meniscus is pulled rapidly through a pore. In a drying soil, it is, therefore, the radii of the pore-necks rather than the radii within the enlarged section of the pore, that are important. In such a system, $r$ in Equation (5) is the radius of the pore neck.

Suppose now that the same pore initially contains water only in the narrow section at the base and that the meniscus is of radius $r_3$. With increasing matric potential, because of introduction of water at the base of the system, the meniscus will gradually rise until it reaches the widest point in the lower section of the pore, where $r = r_4$. Here the system becomes unstable because the pore immediately ahead is of radius less than $r_4$. The meniscus, therefore, moves rapidly through to the position where $r = r_5(= r_4)$. Here again, the radius of the meniscus will confer equilibrium with the potential. A further rise in potential will cause the meniscus to rise to $r = r_1$ and beyond. In this example, where wetting rather than draining is occurring, water enters the system in gushes, but the critical radii ($r$ in Equation (5)) and potentials relate to the widest portions of each pore segment rather than to the pore necks.

This different behaviour of water during drying (desorption) and wetting (adsorption) means that there is no one relationship between soil water content and potential and the relationship is further affected by irreversible changes in the arrangement of soil particles. These changes are often associated with change in water content. Thus, soils exhibit hysteresis in the relationship between water content and matric potential. This topic is more fully discussed by Youngs (1965).

The relationship between matric potential and water content of a soil varies with the frequency distribution of the various pore sizes and so with soil type. The relationship is conveniently expressed in the graphical form called the 'moisture characteristic curve' which is a particular case of the general water sorption isotherm. The hysteresis phenomenon gives rise to a family of curves for each soil and in Fig. 5.4 curves are given for soils drying from saturation and for the same soils wetting from the oven-dry condition. These are the boundary curves of a hysteresis envelope and innumerable curves for the same soil lie within this envelope. Each curve is appropriate to a given history of partial wetting and drying. It is, therefore, not possible to infer a matric potential from measurements of water content for an experimental soil, wetted from the air-dry condition, from the moisture characteristic curve for the same soil drying from saturation. The error introduced by hysteresis may be many bars potential, a fact unfortunately not recognized by many workers. The moisture characteristic must be produced for a soil treated in precisely the same way as the soil in the actual experiments.

The terms 'field capacity' and 'permanent wilting point' frequently occur in the literature. The first refers to the water content of a soil, draining from saturation, when the rate of water loss due to gravity becomes small. The field capacity of a soil is, therefore, not a precise measure but the corresponding matric potentials are frequently in the range −0·1 to −0·5 bar (Colman, 1947). The permanent wilting point of mesophytic higher plants is affected by hydraulic conductivity factors as well as by matric

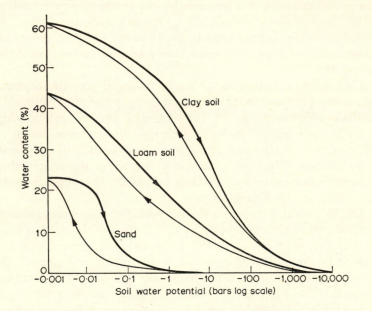

Fig. 5.4. *Moisture characteristic curves for three soils. The curves shown are the drying (upper) and wetting (lower) boundary curves.* (After Yong and Warkentin, 1966)

potential and also varies from species to species (Newman, 1969a, b). Like field capacity, it thus has a somewhat arbitrary value, taken as about −15 bar (Slatyer, 1967). Neither field capacity nor permanent wilting point are concepts of great direct importance in soil microbiology although it should be remembered that soils with normal drainage will be wetter than field capacity for only short periods. Surface soils, however, may be far drier than the permanent wilting point even though plants show no sign of stress because they are still supplied with water from deeper horizons.

### 5.3.3. Osmotic potential

The potential of soil water can be decreased not only by the effects of interfaces on matric potential but also by the presence of solutes within it. The presence of such solutes decreases entropy because some of the water molecules adopt a more ordered arrangement around the solute molecules. Traditionally, this decrease in entropy is thought of in terms of osmotic potential, which may be measured in the same units as matric potential. Solute concentration may be accurately related to osmotic potential ($\pi$, bar) by the equation:

$$\pi = -RT\rho vm\phi/10^9$$
$$= -24 \cdot 7vm\phi \text{ (at 25°C)} \tag{6}$$

where $R$ is the gas constant ($8 \cdot 31 \times 10^7$); $T$, the temperature (K); $\rho$, the density of water at $T°$; $v$, the ions molecule$^{-1}$ (taken as 1 for nonionic solutes); $m$, the molality; $\phi$ the osmotic coefficient at molality $m$ and $T°$. The osmotic coefficients of many solutes are given by Robinson and Stokes (1955). It should be carefully noted that solute concentrations must always be expressed in molality (g solute per 1,000 g solvent) and not in molarity (g solute per 1,000 cc solution).

It is often considered that the osmotic potential of a solution is effective only in the presence of a semipermeable membrane and this is, of course, true if effectiveness of osmotic *pressure* in moving water is concerned. In the presence of a semipermeable membrane, water moves from zones of high to zones of low osmotic potential but water movement is negligible in the absence of a membrane: then the solute molecules redistribute themselves and so equalize the potential. Microbiologically, however, it is by no means obvious that a semipermeable membrane is necessary for the major effect of osmotic *potential* to be felt. Imagine a cell, bounded by a semipermeable membrane, to be placed in a hypertonic solution. The cell will lose water to the exterior until its cytoplasm becomes isotonic. If the cell is to metabolize, its enzymes must be capable of functioning in the decreased potential now existing in the cytoplasm. Consider now another cell bounded by a membrane that is permeable to the external solute. As the solute concentration is greater outside than inside the cell, the solute will diffuse inwards until equilibrium is reached. The second cell has not lost water but the potential has decreased to the same extent as in the first cell because of the influx of solute. In both cells the degree of hydration of an enzyme at the resultant potential depends upon its ability to compete with the solute for the water molecules present. Skujins and McLaren (1967) have clearly shown that the change in rate of reaction of urease with change in potential follows closely the water sorption isotherm of the enzyme (Fig. 5.5). Activity

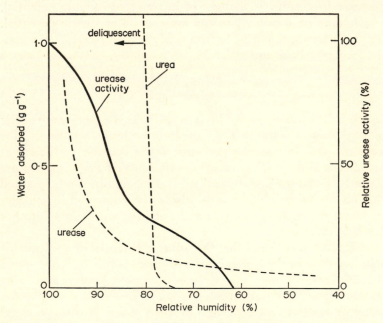

Fig. 5.5. *Urease activity at 20°C (solid line) compared to the water sorption isotherms of the substrate and the enzyme (broken lines).* (After Skujins and McLaren, 1967)

seems to depend upon the degree of hydration of the enzyme (not that of the substrate nor the cytoplasm in general) and this, in turn, depends on potential, regardless of the manner of its imposition.

### 5.3.4. *Total potential; water potential*

Within a soil, water is subjected to many forces, some originating from external gas pressure and the gravitational field. The 'total potential' of soil water is thus the sum of matric, osmotic, gravitational and pneumatic potentials. The sum of the osmotic and matric potentials has been termed the 'moisture potential' or 'water potential' by a number of workers, following Day (1942) and Bodman and Day (1943), and the term 'water potential' will be adopted in this book. In soil microbiology, the gravitational and pneumatic potentials relevant to the system are normally zero so that the total potential and the water potential become the same and some workers have used the terms interchangeably. In most problems, it is the water potential that is particularly relevant, although one or other of the components often predominates to the exclusion of the other (Williams, 1968).

Because the means whereby potential is reduced is relatively unimportant, fungi growing in soil at a given reduced potential are likely to be similar to those growing at the same potential in different systems, such as in stored grain (Christensen and Kaufmann, 1965), high-moisture prunes (Pitt and Christian, 1968), book bindings (Groom and Panniset, 1933), textiles, salted fish and jam (Ingram, 1957; Scott, 1957).

### 5.3.5. *Solute diffusion*

As the water content of a soil decreases, the pathways through which solutes can diffuse from areas of high to areas of low concentration also decrease. As a first approximation, the rate of transfer per unit area of cross-section of soil ($dq/dt$) is given by

$$dq/dt = kD\theta \, dc/dx \qquad (7)$$

where $D$ is the diffusion coefficient of the solute in water; $\theta$, the volumetric water content of soil; $dc/dx$, the concentration gradient; and $k$ is a tortuoisity factor (by analogy with gaseous diffusion). The diffusion of ions or polar molecules is greatly modified by the close proximity of charged surfaces such as those of clay particles, and the effect of soil water content on their diffusion is complex (Calvet 1967a, b; Viets, 1967). For unionized and non-polar molecules, however, Equation (7) is of value and the rate of transfer in a given soil is approximately proportional to the volumetric water content.

### 5.3.6. *Movement of water within soil*

Because of their small size and the negligible loss of water from their surfaces, the rate of absorption of water by soil microorganisms is small compared to that of higher plants (Griffin, 1963a).

G

Uptake of water by microorganisms, therefore, causes little disturbance to the distribution of water in soil and so the hydraulic conductivity of a soil is of little direct significance in microbiology. Indirectly, however, it is of great significance. Conductivity falls off rapidly as water content declines so that the rate of movement, and thus of redistribution, of water becomes clearly retarded if the water content is less than about 40% for a clay, 30% for a loam or 25% for a sand. If evaporation occurs from a soil until it becomes drier than these values, it is impossible to restore the soil uniformly to its original water content by simply adding water at one or a few locations. Furthermore, soil water heterogeneity is an outstanding and largely unsolved difficulty in studying microbiological activity adjacent to transpiring higher plants.

In addition to movement of water through soil in the liquid phase, vapour diffusion also occurs. Such movement in the vapour phase, and associated potential differences, may well be important in the activity of fungi in relatively dry soils. As matric potential decreases below −15 bar, water movement in the vapour phase assumes increasing importance (Rose, 1968). Vapour diffusion is particularly important in rather dry soils subject to marked surface cooling and may lead to great differences in potential over small distances as field temperatures change.

### 5.4. Methods of controlling or measuring potential

Methods of measuring and controlling water potentials in soils and elsewhere have been discussed by Scott (1957) and Holmes, Taylor and Richards (1967). Here, the general principles of only six techniques, chosen because of their usefulness in soil mycology and their illustrative value, will be discussed.

#### 5.4.1. *The tensiometer*

A simple tensiometer, shown in Fig. 5.6 in a form sometimes known as Haines' apparatus, consists of a Büchner funnel with a fine sintered-glass base plate and a hanging column of water. (If considerable reduction in potential is required, mercury may replace water in part of the column.) The hanging column is adjustable so that the matric potential of water in the soil can be controlled, because:

$$\tau = -h_w - h_m \rho_m \text{ (cm water)} \tag{8}$$

where $h_w$ and $h_m$ have the significance shown in Fig. 5.6 and $\rho_m$ is the density of mercury. In obtaining the drying boundary curve of the moisture characteristic, a measured weight or volume of soil is placed in the funnel. The soil is then saturated and the column adjusted so that $\tau = 0$. Successive decreases in potential then

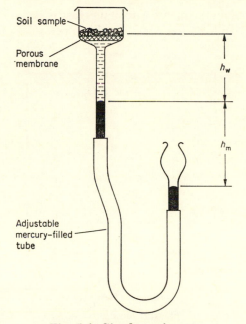

Soil sample

Porous membrane

$h_w$

$h_m$

Adjustable mercury-filled tube

Fig. 5.6. *Simple tensiometer.*
(After Rose, 1966)

withdraw water from the soil. The volume of water withdrawn can easily be measured if the reservoir at the end of the tube is replaced by a burette. If the wetting boundary curve is to be produced, starting with a dry soil, the procedure is tedious because inconveniently long times must be allowed for equilibrium to be reached after each change in potential. Examples of the use of a simple tensiometer in mycology are provided by Griffin (1963*b*, *c*). Rose (1966) has described a more elaborate apparatus, incorporating a suction plate and bubble towers and this has also been used in soil microbiology.

Tensiometers fail if air is drawn through the pores of the porous plate and few withstand potentials below −750 cm water. In any case, a theoretical limit is reached at −1 atm potential when a vacuum develops in the column. Alternative methods are, therefore, necessary if potentials less than about −0·6 bar are to be investigated.

### 5.4.2. *Pressure membrane apparatus*

If soil is placed on the membrane of a pressure membrane apparatus (Fig. 5.7) and the gas pressure about the soil is increased to

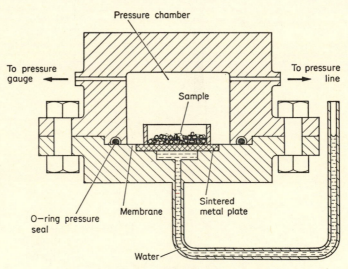

Fig. 5.7. *Pressure membrane apparatus.*
(After Rose, 1966)

$\tau$ atm above atmospheric, then the water content will be that of the soil at −$\tau$ atm matric potential. With apparatus of suitable design, matric potentials within at least the range from −0·2 to −50 bar can be obtained. It should be noted, however, that if air is used as the compressing gas, there will be a partial pressure of 3 atm oxygen if the total pressure is 15 atm. Such a gaseous environment greatly restricts fungal growth (Caldwell, 1963) and it is, therefore, best to remove the soil sample from the apparatus after equilibration and before conducting an experiment. If the experiment is to be performed under pressure within the apparatus, then the composition of the gas should probably be adjusted so that 0·21 atm partial pressure of oxygen is present under the specific operating conditions. Such a procedure will rarely be convenient.

An example of the use of the pressure membrane apparatus in microbiological work is afforded by Kouyeas and Balis (1968) who used it to investigate the effect of matric potential on the restoration of soil fungistasis.

### 5.4.3. *Isopiestic equilibration*

If soil or any other substrate is placed close to a solution, all being within a closed vessel at one temperature, then the vapour pressure of water over the substrate will, in time, equal that over the solution. By extension, the water potential of the substrate will then be the same as the osmotic potential of the solution. The technique, therefore, relies for its accuracy on the availability of precise measurements of the osmotic potential of known solutions. Osmotic coefficients of, or equilibrium relative humidities over, the aqueous solutions of various solutes are available (Robinson and Stokes, 1955; Winston and Bates, 1960; Lang, 1967) so that the relevant potentials can be calculated using Equations (2) or (6). It must be emphasized, however, that it is now the water potential, and not the matric or osmotic potential, that is being determined within the soil.

The successful use of the isopiestic technique depends upon attainment of equilibrium of potential between the substrate and the controlling solution. Equilibration can be promoted by evacuating the vessel, and hence increasing the mean free path of the water molecules, and by maintaining good thermal contact between the substrate and solution. This can be done by placing both in cavities in the same metal block. Winston and Bates (1960) have suggested that the volume of the vessel should not exceed one litre if equilibration is to rely entirely on vapour diffusion. Temperature control is exceptionally important (Schein, 1964), particularly at high relative humidities. It is doubtful if equilibrium can be maintained isopiestically in most biological experiments if the relative humidity exceeds about 95%.

The isopiestic principle has been discussed in relation to microbiology by Scott (1957) and Harris, Gardner, Adebayo and Sommers (1970) and has frequently been applied. Recent examples involving soil fungi are Griffin (1963c, 1966c), Chen and Griffin (1966a, b), Bruehl and Lai (1968), Burgess and Griffin (1968a) and Sommers, Harris, Dalton and Gardner (1970).

### 5.4.4. *Growth in nutrient solutions*

By the use of isopiestic techniques, it is possible to determine the osmotic potential of complex solutions that permit the sustained growth of microorganisms. Although few such determinations have been made, there are sufficient (Scott, 1953; Van den Berg and Lentz, 1968; Sommers, Harris, Dalton and Gardner, 1970) to permit the investigation of the effects of osmotic potential on the growth of microorganisms. If to a nutrient solution of known osmotic potential a further solute is added of known osmotic effect (see references given in previous method), the total osmotic potential of the resulting solution is the sum of the parts. Thus, it is possible to start with a basal medium of, say, −1·2 bar and to adjust the osmotic potential over a wide range by the addition of known weights of solute. In order to make sure that the sole effect of the supplementary solute is an osmotic one, a number of such solutes should be used in the one series of experiments so that stimulation or inhibition caused by any individual solute can be detected. For example, sucrose and other carbohydrates may be metabolized and thus be stimulatory, whereas high concentrations of some ions may be toxic.

With due allowance for specific solute effects, accurate data can be obtained on the effects of osmotic potential by growing fungi in such solutions or on agar media made from them. A good example is provided by a study of *Rhizoctonia solani* (Dubé, Dodman and Flentje, 1971). Plate 2 presents a few of the treatments in another experiment in which *Fusarium culmorum* was grown on a synthetic nutrient agar (−1·2 bar osmotic potential) to which sodium chloride, potassium chloride or sucrose were added to reduce potential. Somewhat unexpectedly, reduction of potential to −8·1 bar both stimulated growth and increased the homogeneity of the colony.

### 5.4.5. *Thermocouple psychrometer*

When a current of about 4 mA is passed in the appropriate direction through a thermocouple junction, that junction is cooled by the Peltier effect (Spanner, 1951; Richards, 1969). If the water potential of the surrounding atmosphere (and nearby soil) is 0 to −80 bar, this Peltier cooling is sufficient to cause water to condense on the junction. After about 30 sec, the current is switched off, evaporation occurs, the junction is cooled and a transient voltage is created in the junction. The voltage generated increases in approximate proportion to the decrease of the water potential below zero but accurate calibration is possible. Thermocouple psychrometers based upon this principle are now being used in soil microbiology and plant pathology (Cook and Papendick, 1970a, b) and have the very great advantage of permitting the measurement of water potential over a wide range in a field or experimental soil.

### 5.4.6. *Filter-paper method*

If a filter paper is buried in soil, the amount of water it absorbs is controlled by the matric potential. The technique has been standardized in a simple and convenient form by Fawcett and Collis-George (1967). Whatman No. 42 filter papers are washed in 0·005% mercuric chloride solution (to reduce microbial growth) and then dried at 105°C. Half of the soil sample is placed in an

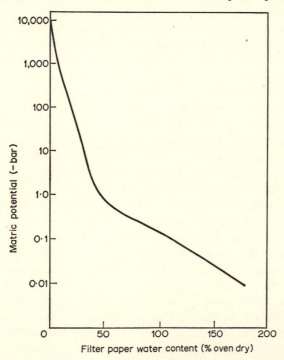

Fig. 5.8. *Wetting characteristic of Whatman No. 42 filter paper.* (After Fawcett and Collis-George, 1967)

air-tight container and three air-dry filter papers are placed on the soil surface. The remainder of the soil is then added and firmly pressed down and the sealed containers are kept at constant temperature in a lagged box. After six or seven days, the centre filter paper is carefully recovered, its water content determined and Fig. 5.8 used to find the corresponding water potential.

### 5.5. Potential and the growth of microorganisms in pure culture

In theory, there is no reason to suppose that potential *per se* will affect microorganisms differently depending upon whether it is altered by osmotic or matric means. There may, of course, be great differences due to other factors associated with either osmotic or matric potentials but these are not direct effects of potential itself. Indeed, if an organism shows a differential response to osmotic and matric potentials of the same magnitude, an interfering associated factor should immediately be suspected. One such factor rarely taken into account is the microbial production of water during the breakdown of some substrates. Thus *Serpula lacrimans* is said to transform over one-third of the weight of a piece of wood into water during decomposition (Miller and Meyer, quoted by Brown, Fahim and Hutchinson, 1968).

Of the techniques for controlling potentials below −0·75 bar, the isopiestic and that in which the organism is in direct contact with a controlling solution, are capable of the greatest precision. The isopiestic technique, however, demands great temperature control for accuracy and is bedevilled by uncertainty as to when exact equilibrium is reached. Even when there is no detectable change in weight of substrate with time, residual potential gradients may exist within the substrate itself. Thus, Sommers, Harris, Dalton and Gardner (1970) report that such gradients exist within thin agar films even after a prolonged period, and are of sufficient magnitude to reduce the accuracy of microbiological experiments. Data from experiments where the organism grows within a nutrient broth or on a nutrient agar of known potentials, are thus likely to be the most accurate if care is taken to minimize the effects of specific solutes. In this section, data obtained with the direct solute technique have been selected wherever possible, but frequently data obtained in other, probably less accurate, ways have had to be used.

In considering the effects of water potential on microorganisms, the growth of fungi will be divided, somewhat arbitrarily into spore germination, vegetative growth and reproduction (sexual or asexual), that of bacteria will be more briefly considered alongside vegetative growth of fungi.

### 5.5.1. *Germination of fungal spores*

Germination has been defined in a number of ways but here I shall use the term to encompass all those stages from the first physiological change marking the break of dormancy to the appearance of a germ tube. The period involved has been called the 'latent period of germination' or the 'activation time' and during it the spore undergoes many changes (Sussman, 1966; Fletcher and Morton, 1970). Prominent among these, for many fungi, is a swelling of the spore and this is accompanied by enhanced respiration and intense metabolic activity (Marchant and White, 1966, 1967; Gottlieb and Tripathi, 1968; Martín and Nicolás, 1970). The rate of respiration is dependent upon water potential (Mozumder and Caroselli, 1970). The relationship between swelling and potentials is, however, unclear on the present evidence (Ekundayo and Carlile, 1964; Ekundayo, 1966; Barnes and Parker, 1966, 1967) because the potentials of some parts of the systems used are not known.

Although the effect of potential on the component parts of the germination process is poorly understood, it has long been known that decreasing potential increases the length of the latent period until, at some potential, germination does not occur. Representative data are given in Table 5.2: more have been given by Griffin (1963a) and Pitt and Christian (1968). The limitations of such data should be noted. An examination of those presented in Table 5.2 for *Aspergillus flavus*, *Phycomyces nitens* and *Rhizopus stolonifer* shows that the minimum potential at which germination occurs tends to decrease asymptotically the longer the experiment is continued. Furthermore, the minimum potential permitting germination increases if nutrient concentration is low (Tomkins, 1929) and is affected by other environmental factors, such as $p$H (Schelhorn, 1950) and temperature (Fig. 5.9). The latent period also increases dramatically with age of spores (Groom and Panniset, 1933; Heintzeler, 1939).

Table 5.2. *Minimum water potential for spore germination and corresponding latent periods*

| Fungus | Potential (-bar) | Latent period | Author |
|---|---|---|---|
| *Aspergillus amstelodami* | 410 | 63 days | Pitt and Christian (1968) |
| *A. flavus* | 108 | 10 days | Pitt and Christian |
| *A. flavus* | 306 | 4 months | Armolik and Dickson (1956) |
| *A. niger* | 234 | 10 days | Pitt and Christian |
| *A. ruber* | 490 | 4 months | Snow (1949) |
| *A. versicolar* | 306 | 63 days | Pitt and Christian |
| *Botrytis cinerea* | 98 | 2 days | Snow |
| *Chrysosporium fastidium* | 516 | 48 days | Pitt and Christian |
| *Cladosporium herbarum* | 174 | 7 days | Snow |
| *Fomes annosus* | 98 | 9 weeks | Rishbeth (1951) |
| *Fusarium culmorum* | 128 | 8 weeks | Schneider (1954) |
| *F. oxysporum* | 157 | 8 weeks | Schneider |
| *Mucor spinosus* | 98 | 3 days | Snow |
| *Paecilomyces varioti* | 234 | 9 days | Pitt and Christian |
| *Penicillium chrysogenum* | 306 | 2 weeks | Galloway (1935) |
| *P. expansum* | 219 | 2 weeks | Galloway |
| *P. fellutanum* | 306 | 28 days | Pitt and Christian |
| *Phycomyces nitens* | 70 | 2 days | Walter (1924) |
| *P. nitens* | 204 | 15 days | Heintzeler (1939) |
| *Rhizopus stolonifer* | 98 | 2 days | Snow |
| *R. stolonifer* | 237 | 16 days | Heintzeler |
| *Sclerotium rolfsii* (sclerotia) | 14 | 14 days | Abeygunawardena and Wood (1957) |
| *Scopulariopsis brevicaulis* | 145 | 2 weeks | Galloway |
| *Verticillium alboatrum* | 115 | 25 days | Mozumder and Caroselli (1966) |
| *Xeromyces bisporus* | 690 | 120 days | Pitt and Christian |

## 5.5.2. *Vegetative growth of fungi and bacteria*

Although it is possible that the most rapid germination of spores might occur at zero potential, this cannot be so when vegetative growth is considered. Exogenous nutrients are necessary for growth

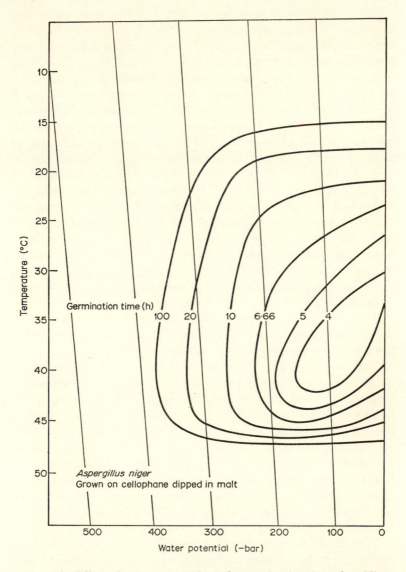

Fig. 5.9. *Effect of water potential on the germination time of conidia of* Aspergillus niger. (After Bonner, 1948)

and these must reduce the potential osmotically, even if only to a small degree. Thus, organisms are likely to have evolved so that their most rapid growth occurs at potentials commonly found in their substrates. Such an optimum at somewhat reduced water potentials has now been shown for a number of fungi. It is of interest that this phenomenon can best be shown when the fungi are grown in direct contact with the controlling solutes and that it it is more difficult or even impossible to demonstrate when potential is controlled by other techniques. In the latter case, the phenomenon is almost certainly masked by interfering factors, such as reduced solute diffusion in drier substrates (Sommers, Harris, Dalton & Gardner, 1970). In Fig. 5.10 the effect of water potential on the linear growth rate of hyphae of representative fungi is illustrated. In the lower potential range for each fungus, the curves become asymptotic and, in Fig. 5.10, have not been shown for in this range the growth rates are negligible compared with the maximum. Data on the effects of water potential on dry weights of fungi are not available. It is clear that fungi differ greatly among themselves in regard to response to potential, some being reduced to negligible activity at $-40$ bar, other growing at half their maximal rate at $-100$ bar. Some, such as *Xeromyces bisporus* and strains within the *Aspergillus glaucus* group (Griffin, 1963d), are unable to grow at high potentials and are exceptional in their ability to grow at very low potentials. Whereas *X. bisporus* and *Saccharomyces rouxii* (with the lowest known minimum potentials for growth of $-690$ and $-693$ bar, respectively (Pitt and Christian, 1968; Schelhorn, 1950)), have not been recorded from soil, members of the *A. glaucus* group are common there. Most fungi, however, are restricted, so far as reasonably rapid growth is concerned, to potentials exceeding $-60$ or $-80$ bar. Growth, like germination, is affected by temperature and nutrient concentration and occurs at lower potentials if temperature is optimum and nutrients are not limiting (Tomkins, 1929; Snow, 1949; Ritchie, 1957, 1959; Te Strake, 1959).

The effect of water potential on the growth of soil bacteria has not been investigated accurately. In general, however, bacteria are limited to potentials greater than $-100$ bar although a few grow at about $-200$ bar (Scott, 1957; Wodzinski and Frazier, 1960, 1961a, b; Christian and Waltho, 1962; Lanigan, 1963; Strong, Foster and Duncan, 1970). The growth of soil strepto-

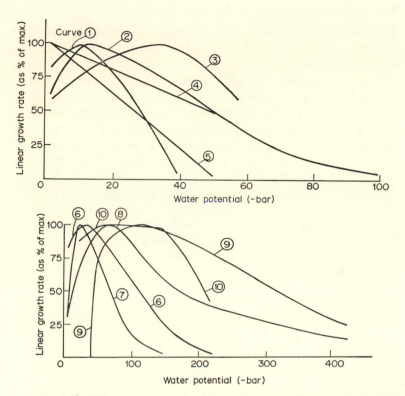

Fig. 5.10. *Effect of water potential on the rate of linear extension of eleven fungi*

Curve 1.  Phycomyces nitens (*J. M. Wilson, unpublished data*) *and* Phytophthora cinnamomi (*Sommers, Harris, Dalton and Gardner, 1970*).

Curve 2.  Fusarium culmorum (*isolate from Washington State, U.S.A. – personal unpublished data*).

Curve 3.  Fusarium culmorum (*isolate from New South Wales, Australia – personal unpublished data*).

Curve 4.  Cochliobolus sativus (*personal unpublished data*).

Curve 5.  Ophiobolus graminis (*personal unpublished data*) *and* Rhizoctonia solani (*Dubé, Dodman and Flentje, 1971*).

Curve 6.  Aspergillus niger (*Heintzeler, 1939*).

Curve 7.  Aspergillus flavus (*Ritchie, 1959; Pitt and Christian, 1968*).

Curve 8.  Aspergillus amstelodami (*Scott, 1957*).

Curve 9.  Xeromyces bisporus (*Scott, 1957*).

Curve 10. Stereum frustulosum (*Bavendamm and Reichelt, 1938*).

mycetes on agars of different osmotic potentials becomes negligible at −80 bar and is greatly reduced at −40 bar (T. W. Wong, unpublished data).

Evidence exists for the active transport of water against potential gradients in some animals (Andrewartha, 1964; Beament, 1965) but I know of no conclusive positive evidence for fungi, although such an ability has been suggested (Ekundayo and Carlile, 1964; Park and Robinson, 1966). Rather, the data indicate that for fungi (Table 5.3), as for bacteria (Christian and Waltho, 1962, 1964), the

Table 5.3. *Approximate osmotic potentials of hyphae and their substrates*

| | Osmotic potential (−bar) | | |
| | Substrate | Apical cells | Author |
| Fungus | | | |
| --- | --- | --- | --- |
| *Fusarium oxysporum* | 3·0 | 5·6 | Robertson (1958) |
| *Aspergillus niger* | 0·3 | 3·6 | Park and Robinson (1966) |
| *Neurospora crassa* | 3·0 | 18·0 | Robertson and Rizvi (1968) |

cell contents have an osmotic potential lower than that of the medium. Similarly, the osmotic potentials of the ascus and ascospores of *Sordaria fimicola*, grown on a complex medium unlikely to have a potential less than about −5 bar, were reported to be within the ranges of −10 to −30 bar and −22 to −45 bar, respectively (Ingold, 1966; Milburn, 1970). The only data on the turgor pressure of hyphal cells (virtually a positive hydraulic potential) are those for *Neurospora crassa*, indicating a value of about 15 bar (Robertson and Rizvi, 1968). Unfortunately, all these data are for fungi grown on substrates of high potential and they shed little light on the internal potentials of hyphae exposed to low potentials.

It has been noted that some fungi grow only at reduced potentials. This limitation appears to be a direct response to potential and is not akin to the situation associated with halophilic bacteria, which require high concentrations of specific ions for their cellular integrity and metabolism (Ingram, 1957; Larsen, 1967). The only exceptions are for certain abnormal strains of fungi (Siegel, Roberts, Lederman and Daly, 1967) and, more interestingly, for the marine fungus *Thraustochytrium roseum*. This fungus exhibits both a positive response to reduced potential and specific ion stimulation. Thus, an increase in the rate of endogenous respiration in the presence of 0·2 to 0·4 molar sodium chloride is primarily an osmotic effect, whereas enhanced phosphate transport is specific to the sodium ion (Siegenthaler, Belsky and Goldstein, 1967). It is possible that organisms with similar responses occur in saline soils. I know of no work on the physiology of fungal isolates from saline soils but the species isolated from salt marshes are common soil fungi (Pugh, 1962). Soil fungi isolated on agars of low osmotic potentials are the same species (mainly of *Aspergillus* and *Penicillium*) as those that are active in other situations of physiological drought (Chen, 1964). High salinity in the surface soil is characteristic of areas in which *Coccidioides immitis* occurs (Elconin, Egeberg and Egeberg, 1964). Such a distribution for this human pathogen is in accord with its tolerance, relative to some antagonists, of high salinity combined with high temperature (Egeberg, Elconin and Egeberg, 1964).

Growth at very low potentials poses difficulties because of the dependence of enzymic activity on water (Acker, 1962). Of particular interest is the demonstration by Falk, Hartman and Lord (1962, 1963*a*, *b*) that the DNA molecule is highly hydrated until potential decreases to −115 bar and that with further reduction water molecules are progressively removed. Between about −400 and −800 bar the helical structure is reversibly distorted and the base pairs become disordered. *Xeromyces bisporus*, growing at −690 bar, must thus be exploiting almost the total range of water potential theoretically available to an organism in which DNA activity is essential for growth. Wilson and Harris (1968) have shown that some phosphorylation occurs even at −880 bar but that incorporation of inorganic phosphates into adenosine triphosphate, uridine diphosphate hexose and nicotinamide adenine dinucleotide is, at best, extremely slow at −130 bar. The slow growth of microorganisms at potentials approaching −100 bar may thus be associated with the slow formation of high energy phosphate centres.

## 5.5.3. *Reproduction*

Pitt and Christian (1968) concluded that, although the reported minimum potential for germination and growth of ascomycetes and related imperfect fungi was often lower than that for asexual sporulation, there was no evidence of an intrinsic inability to sporulate at the lower potential. Rather, they attributed the difference to a premature termination of experiments, because they had noted that for a given species and regardless of potential, asexual reproduction commenced only when the colony had reached a certain critical size. Conditions for sexual reproduction in ascomycetes were thought to be more exacting, regardless of the duration of the experiment, and appeared to require up to 100 bar higher potential than that required for germination. The few data for phycomycetous fungi (Heintzeler, 1939) suggest that their asexual reproduction may be more sensitive to potential.

## 5.6. Movement of microorganisms in soil water

In the last section, we considered the direct effect of water potential or its components upon the growth of microorganisms and found the effect on bacteria and fungi to be similar except for a few unusual organisms. Concurrent changes in matric potential and water content of a particulate system are, however, likely to have a different affect on filamentous organisms, such as fungi and actinomycetes, and on unicellular organisms or structures, such as bacteria, protozoa and fungal zoospores. It must be emphasized that these expected differences are at the moment based largely on theoretical considerations. Supporting experimental data are often circumstantial and qualitative but the implications are so great that this topic will, I believe, come to have profound significance in soil microbiology.

A simple introduction to the topic of microbial movement is provided by the growth of fungi and bacteria on agar media. If such a medium is allowed to stand after being poured into a Petri dish, the thickness of the water film over its surface is only a small fraction of a micron (at $-0.33$ bar matric potential, the thickness of a water film on a solid flat surface is only about 20 molecules). Mycelium is able to spread rapidly over and beneath such a surface because the older parts of the hypha provide an anchorage on the surface while growth is localized at the apex of a relatively rigid and approximately linear hypha. The substrate is thus rapidly explored and utilized. Bacteria, however, remain as small, localized colonies on such a surface. Reproduction leads to the formation of another simple cell that is usually pulled back within the boundary of the parent colony by surface tension. Directed linear growth outwards is usually impossible because of the lack of a rigid, elongate form. Bacteria rapidly spread over a surface only in those regions where a water film, comparable in thickness to the size of the bacterium, is present.

In soil, water does not lie on flat surfaces but is held within pores, but the same basic considerations apply. Hyphae will be able to spread along the walls of drained pores, or even across the pore from side to side, whereas bacterial spread will, at best, be extremely slow in the absence of continuous water pathways of the requisite dimensions. If it is accepted that bacterial spread, and hence efficient utilization of a substrate, is bound up with the existence of water-filled pores, then the following considerations apply. First, the smallest diameter along the length of the pore must be large enough to permit passage of the bacterium, whether by Brownian or flagellar movement. Second, whether such a pore and those larger than it, will be air or water-filled will depend upon the matric potential of the system. Third, to permit appreciable movement, there must be enough water-filled pores of the requisite diameter to provide a continuous pathway.

The validity of these limiting conditions have been tested in a study of *Pseudomonas aeruginosa* (Griffin and Quail, 1968). The cells of this bacterium are rod-shaped, $0.5 \times 1.5 - 3.0$ $\mu$m. We argued, therefore, that pore necks of radius less than $1-1.5$ $\mu$m were likely to restrict severely the rate of passage. Such necks would drain at $-1$ to $-1.5$ bar matric potential say, for the sake of simplicity, at $-1$ bar. Therefore, all water held in pores which would drain only at potentials below $-1$ bar may be disregarded as being unavailable to act as bacterial pathways. The second limiting condition is that associated with continuity: how much greater water content than that of a soil at $-1$ bar potential is necessary to provide the necessary continuity of pathways? I know of no way to arrive at a conclusion theoretically but in a number of systems it was shown experimentally for *P. aeruginosa* to be about $0.11$ cm$^3$ water cm$^{-3}$ soil. Some of the consequences of this are best appreciated by reference to Fig. 5.11.

H

The choice of *P. aeruginosa* in this early work was fortuitously favourable, because subsequent experiments with other bacteria, mainly *Bacillus* spp., have yielded far less clear-cut results (T. W. Wong, unpublished data). In part, this is probably due to a dependence of cell size and motility on the nutrient concentration in the water. Such differences make it difficult to evaluate the limiting conditions and there is little doubt that the work with *P. aeruginosa* led to the development of an unduly simplified picture. In particular, for small bacteria, only water held below −6 bar matric potential (rather than −1 bar) should probably be dis-

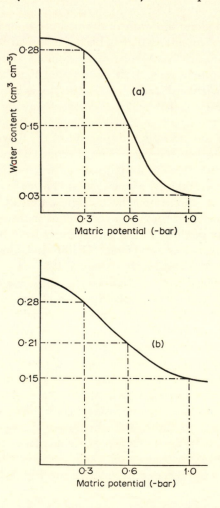

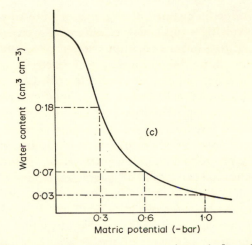

Fig. 5.11. *Movement of* Pseudomonas aeruginosa *in hypothetical soils.* (Based on Griffin and Quail, 1968; Griffin, 1970)

Explanation: *In the loamy sand* (a), *water content at* −1 *bar matric potential is 0·03 cm³ cm⁻³, and this water may be disregarded as being in pores too small for rapid movement of this bacterium. The 'base level' for water content may thus be taken as 0·03 cm³ cm⁻³. At* −0·3 *and* −0·6 *bar potential, water content is 0·28 and 0·15 cm³ cm⁻³, respectively, and at both potentials bacterial movement over appreciable distances will occur because the water contents are more than the critical 0·11 cm³ cm⁻³ greater than the 'base level'. By analogous argument, movement will occur in the clay loam* (b) *and in the sand* (c) *at* −0·3 *bar potential but not at* −0·6 *bar, because in neither case at* −0·6 *bar does the water content exceed the relevant 'base level' by the critical amount.*

regarded. None the less, the basic considerations appear to be valid and have important consequences.

It has sometimes been argued that in soil, bacteria are literally ubiquitous and that there is, therefore, no need for them to move appreciably to colonize a substrate within the soil. Such an argument is fallacious in the light of data on bacterial distribution (Gray, Baxby, Hill and Goodfellow, 1968). First, it has been estimated that less than 0·1% of the total available surface area within a soil is colonized by bacteria at any given moment. Second, bacteria are not uniformly distributed but usually occur as discrete colonies of one species, each with from a few to a hundred or more cells. Thus if a series of degradations are to occur at one microsite, involving a succession of different species (Stotzky and Norman, 1961a), movement will be necessary. Another argument tending to minimize the importance of bacterial movement is based on the supposition that nutrients will diffuse in the soil water to a sedentary cell. Solute diffusion, however, decreases linearly, as a first approximation, with volumetric water content so that it is restricted at those lower potentials that also restrict bacterial movement. Furthermore, if enzymic hydrolysis of the substrate is a pre-requisite, many enzymes are bound to the surfaces of bacteria and those that are soluble are strongly absorbed by clays and colloids. All these factors point to the necessity, convincingly argued by Greenwood (1968a) and Russell (1968), for bacteria to come very near to their substrate, either by their own movement or by movement of the substrate through growth, tillage or natural soil movement.

Regardless of the direct effect of potential on growth, bacterial activity is, therefore, likely to be reduced rapidly as matric potential falls because of decreasing ability to move on to or over the substrate. As we have seen, the critical range in most natural soils is likely to be $-0.5$ to $-5$ bar. At potentials below these, the postulated decline in activity will reduce the bacterial component of soil respiration and the antagonistic activity of bacteria. Because of their different morphology, there is no reason to believe that fungi will be affected in the same way so that the main antagonists of fungi in drier soils are likely to be actinomycetes and other fungi, and in wetter soils, bacteria (Finstein and Alexander, 1962). Circumstantial evidence on many of these points will be noted in the following section.

This argument concerning the limiting conditions for movement of bacteria applies equally to other unicellular structures, such as protozoans and zoospores. Because of their relatively large size, it is unlikely that they are able to move far except in large pores of radius 5–10 $\mu$m, draining at potentials of $-0.2$ to $-0.3$ bar. Such pores will be water filled only in very wet soils, in most cases during or immediately after heavy rain. Furthermore, motility of zoospores has been shown to be decreased greatly by frequent contact with solid surfaces (Ho and Hickman, 1967; Bimpong and Clark, 1970) such as must occur in pores. In these conditions passive movement of the zoospores in moving water may be far more important than movement brought about by flagella (Hickman and Ho, 1966; Dick, 1968; Griffin, 1969). Infection of strawberry by *Phytophthora fragariae* increases with increasing soil water content, increasing coarseness of soil texture and with soil water movement (Hickman and English, 1951), thus supporting the general hypothesis.

The more general importance of spore dispersal within a soil brought about by percolating water is uncertain. Earlier, I reviewed evidence, provided mainly by Burges (1950), Rishbeth (1957), Park (1959) and Hepple (1960), on the movement of spores by soil water (Griffin, 1963*a*). So long as the spores are in a near-saturated soil with strong movement of water, spore transport occurs within limits provided by the relative sizes of spore and pore (Dickinson and Parkinson, 1970). In general, the distribution of spores from a point source is cone shaped with the greatest lateral spread at the level of the source. Temporary drainage impedes subsequent movement, however, presumably because of the adherence of spores to surfaces with which they had come into close contact. Xerospores, with dry walls, do not usually move easily in soil water but this may be of little significance if their unwettable surface properties are short lived in soil (Dobbs and Hinson, 1960).

Most studies have involved relatively uniform soil samples without the natural channels provided by cracks, earthworms and dead roots. As the equivalent of extremely heavy rainfall is needed to move spores through a uniform soil, I have suggested (Griffin, 1963*a*) that the importance of spore movement through large channels may well exceed that through the soil pores. Although soil and spores are certainly washed to a considerable depth by heavy rain falling on to a cracked soil, the significance of movement within earthworm and dead root channels is more problematical. I was fortunate enough to be within the underground root observation laboratory at East Malling Research Station at a period of torrential rain in 1968 when the soil surface became covered by a persistent sheet of water. The earthworm channels remained air-filled, probably because at no place were they continuous with the external water. The heavy rain had broken down the surface soil so that a fine relatively unstructured layer lay between the top of the channels and the overlying water. The rate of movement of water through the soil was thus governed mainly by that through the surface layer and was never high enough for the large channels to fill.

### 5.7. Water potential and the activity of microorganisms in soil

The frequency of bacteria occurring on buried slides and within potato lenticles decreases rapidly as water content of the soil falls below field capacity and the matric potential approaches −1 bar (Rybalkina and Kononenko, 1957; Kouyeas, 1964; Lapwood, 1966; Lewis, 1970). Similarly, the rates of nitrification and sulphur oxidation (both bacterial activities) are maximal between −0·1 and −0·2 bar and −0·03 and −0·06 bar matric potential respectively, and become negligible in most soils at −15 bar (Moser and Olsen, 1953; Dommergues, 1962; Justice and Smith, 1962; Miller and Johnson, 1964; Dubey, 1968; Sabey, 1969). Bacteria also increased far more at 0 and −0·8 bar than at −2·2 bar matric potential when fungi were added as a substrate to soil (Bumbieris and Lloyd, 1967). Bacterial numbers in two soils amended with glucose and ammonium sulphate were greatest at water potentials equal to, or exceeding, −5 bar and declined rapidly with drop in potential below this value. In the same soils, 40–50% of the chlamydospores of *Fusarium culmorum* germinated within 24 hr at potentials down to −60 bar and some germination occurred within 72 hr at −85 bar. As potentials increased above −10 bar, however, progressively more of the germ tubes lysed, or produced new chlamydospores, within 48–72 hr although germ tubes persisted at lower potentials. A relationship between bacterial activity and non-persistence of germ tubes was further supported by the suppression of lysis in soils further amended with streptomycin and neomycin (Cook and Papendick, 1970a).

I consider that these decreases in bacterial activity were primarily the result of decreased mobility, as discussed in the previous section, because the potentials involved were relatively high. The lower limit for bacterial activity in soil imposed directly by water potential, is probably at about −80 bar. Microbiological activity at still lower potentials is probably entirely fungal (Dommergues, 1962).

Within natural soil, far more is known about the activity of individual species of fungi than of bacteria. Kouyeas (1964) has shown that buried plant parts were colonized most frequently by

*Pythium* spp., *Mortierella* spp. and predacious hyphomycetes at water potentials exceeding −1 bar whereas *Fusarium* spp., *Trichoderma viride* and *Gliocladium roseum* were little affected in the range −0·4 to −20 bar. Species of *Mucor*, *Aspergillus* and *Penicillium* were isolated with increasing frequency as the potential decreased.

The activity of fungi in soils at potentials less than those of Kouyeas have been studied by Griffin (1963*d*, 1966*c*) and Chen and Griffin (1966*a*, *b*). In most experiments, the combined activity of the soil mycoflora was assessed in terms of the density of hyphae

Table 5.4. *Density of viable hyphae on hair on the soil surface (on an arbitrary scale, 0 to 5\*) (after Chen and Griffin, 1966a)*

| Water potential (−bar) | Incubation period (days) | | | | | | | | | | | | |
|---|---|---|---|---|---|---|---|---|---|---|---|---|---|
| | 2 | 4 | 7 | 9 | 14 | 21 | 28 | 42 | 56 | 77 | 78 | 119 | 476 |
| *ca.* 15 | 2 | 5 | 5 | 5 | 5 | 4 | 5 | 5 | 3 | 4 | 4 | 4 | — |
| 70 | 1 | 3 | 4 | 5 | 5 | 5 | 5 | 5 | 5 | 5 | 5 | 5 | — |
| 145 | 1 | 1 | 2 | 3 | 5 | 5 | 5 | 5 | 5 | 5 | 5 | 5 | — |
| 219 | 0 | 0 | 0 | 0 | 2 | 2 | 3 | 4 | 5 | 5 | 5 | 5 | — |
| 306 | 0 | 0 | 0 | 0 | 0 | 0 | 0 | 1 | 2 | 2 | 3 | 4 | — |
| 395 | 0 | 0 | 0 | 0 | 0 | 0 | 0 | 0 | 0 | 0 | 0 | 0 | 2 |

\* 0 – no hyphae visible; 3 – all hairs in contact with soil were colonized; 5 – hairs not in direct contact with soil were heavily colonized.

occurring on originally sterile hair fragments 2–4 mm long, dusted on to the soil surface. The time elapsing before visible colonization occurred increased steadily with decreasing potential (Table 5.4) indicating that the rate of decomposition of soil organic matter will be negligible for most purposes at potentials less than −150 bar. The activity of individual species was usually assessed in terms of the frequency of colonization of the hair fragments. A clear ecological differentiation between species was shown to exist, based upon water potential (Fig. 5.12). At potentials less than −145 bar, the active flora consisted almost exclusively of species of *Aspergillus* and *Penicillium*. It is noteworthy that species capable of growth in soil at potentials as low as −300 bar were apparently as prevalent in soils from an English pasture and an

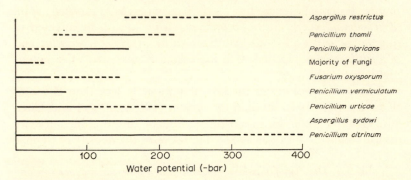

Aspergillus restrictus

Penicillium thomii

Penicillium nigricans

Majority of Fungi

Fusarium oxysporum

Penicillium vermiculatum

Penicillium urticae

Aspergillus sydowi

Penicillium citrinum

Water potential (-bar)

Fig. 5.12. *Relationship between water potential and frequency of isolation of fungi from hair which had been colonized by soil-borne fungi at 25°C (full line – frequent; broken line – less frequent).* (Based largely on data of Chen and Griffin, 1966a)

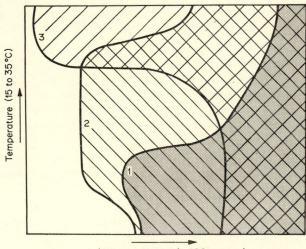

Temperature (15 to 35°C)

Water potential (−400 to 0 bar)

Fig. 5.13. *Diagram of the intersection between temperature (15–35°C) and water potential (0 to —400 bar) in determining the pattern of activity of fungi in colonizing pieces of hair lying on soil. Group 1 includes* Penicillium chrysogenum, P. frequentans; *group 2*, P. citrinum, Aspergillus versicolor; *group 3*, A. flavus, A. niger, A. terreus. (After Chen and Griffin, 1966b)

Australian rainforest as they were from a desert. Thus, no evidence was provided for any selection of a xerophytic population in arid zones. There was a marked interaction between temperature and water potential in determining those species of *Aspergillus* and *Penicillium* recovered from the hair (Fig. 5.13). In general, *Aspergillus* spp. were of equal or greater prevalence relative to *Penicillium* spp. at all potentials at 30° and 35°C and at −220 to −400 bar at 15° and 20°C. This is in accord with the concept of Peyronel (1956) that *Aspergillus* is a more xerothermic genus than *Penicillium*.

The differential effect of low potentials on the activity of many fungi is of general significance, for surface soils become very dry in most parts of the world for at least a few days at a time. In such a situation, the survival of hyphae of *Cephalosporium gramineum* in wheat straw at 15°C was best at the lowest potential tested (−270 bar) and worst at −145 and −200 bar, where *Penicillium* spp. were dominant (Bruehl and Lai, 1968). Similarly, antagonism by *Penicillium* spp. was thought to be the cause of poor survival of *Gibberella zeae* in wheat straw lying on soil at 25°C and −190 and −390 bar potential. Recovery of *G. zeae* was good from straws placed on soil at −1,540 bar potential (when no fungi were active and there was, therefore, no antagonism) (Burgess and Griffin, 1968a). The difference in the temperature of the experiments (15° and 25°C) probably accounts for the different potentials permitting the most antagonism.

In arid and semi-arid regions, the increase of ammoniacal nitrogen in the dry season is probably associated with the decomposition of organic matter at low potentials by fungi, in the absence of uptake by higher plants and nitrification by bacteria (Dommergues, 1962).

In this treatment of microbial activity, the paucity of references to the soil streptomycetes is obvious and is due to an almost total lack of information. Kouyeas (1964) found them to increase with every reduction in potential to the lowest used (about −20 bar) and Chen and Griffin (1966a) observed them to be active only at potentials greater than −55 bar. *Streptomyces scabies* colonized potato lenticels more at −0·8 bar than at −0·13 bar matric potential, probably because of reduced bacterial colonization at the lower potential (Lapwood, 1966; Lewis, 1970). The apparent dominance of streptomycetes in many dry soils, as revealed by dilution plate counts, is likely to be attributable to the resistance of their spores to desiccation rather than to hyphal activity at low potentials.

The numbers of fungal propagules decline in soils subjected to extreme desiccation in such countries as India and Australia (Saksena, 1955; Warcup, 1957) but fluctuations associated with soil moisture are hard to detect in temperate regions. Viability may be lost because of the antagonism of organisms that remain active or because of loss of water. Conversely, survival may be enhanced if the level of endogenous respiration or the tendency to spontaneous germination are reduced.

### 5.7.1. Influence of solute diffusion

The diffusion of solutes in soil water is of general microbiological significance but most of the few published experiments relate to plant pathogens. Thus an association between exudation from seeds and pre-emergent mortality has been demonstrated (Flentje and Saksena, 1964; Schroth and Cook, 1964; Matthews and Whitbread, 1968; Hayman, 1969). Flentje (1964) suggested that the main influence of soil water on mortality might be mediated by changes in the rate of diffusion of materials from the seed to the pathogen. It remained for Kerr (1964) to show, first, that the loss of weight of pea seeds was directly related to the amount of sugars diffusing from them; second, that the percentage of infection of peas by *Pythium ultimum* in each of three soils was directly proportional to their loss of dry weight; and third, that within each, soil infection and loss of dry weight were correlated with soil water content and matric potential. Kerr's argument that the main effect of changes in soil water on infection was brought about by changes in the rate of diffusion of solutes was supported by the finding that increases in soil bulk density at constant gravimetric water content caused increased mortality.

Loss of weight by pea seeds during germination was also shown to be correlated with germination of chlamydospores of *Fusarium solani* f. *pisi*. Both decreased with soil water contents below 8·7% (Cook and Flentje, 1967). That the main factor was solute diffusion was indicated by the uniform germination of chlamydospores over a wide range of soil water content when sucrose and ammonium sulphate were mixed into the soil to obviate the necessity for nutrient diffusion. Germ tubes were extensively lysed within

20 to 48 hr in natural, but not in sterile, soil containing more than 8·7% water. The amount of lysis in these wetter soils was correlated with loss of dry weight of the seeds between 20 and 48 hr and thus, probably, with nutrient diffusion to antagonistic microorganisms during a critical period.

Temperature has a number of effects on the diffusion of sugars and amino acids from plant parts and hence on damping-off by such a fungus as *Rhizoctonia solani*. Increasing temperature must increase the diffusion coefficient of the solutes in water but, within the range from 10° to 30°C, it also reduced the rate of release of solutes from cotton and bean. Above 30°C, exudation increased with rising temperature. Damping-off of cotton declined with temperature rise between 18° and 30°C, suggesting that the predominating effect was on exudation (Schroth, Weinhold and Hayman, 1966; Hunter and Guinn, 1968; Hayman, 1969).

Germination of sclerotia of *Sclerotium cepivorum* in the presence of *Allium* roots also appears to be sensitive to rates of solute diffusion because germination increased throughout the range of soil water content from 20 to 90% saturation (Coley-Smith, 1960). The likeliest explanation is that water content affects the rate of diffusion from the roots of a non-volatile, water soluble precursor which is then decomposed by soil bacteria to yield a volatile, insoluble chemical that stimulates the germination of sclerotia (King and Coley-Smith, 1969).

In the more general sphere, Seidel (1965) showed that soil fungistasis increased in intensity with increases in soil water content. Although he interpreted his results as indicating enhanced diffusion of an inhibitor, they might also indicate enhanced diffusion of nutrients away from the spore, in accord with the hypothesis of Ko and Lockwood (1967). The extensive analysis of Kouyeas and Balis (1968) throws no light on the issue for, in their experiments, the actual determination of spore germination and hence fungistasis were made with all the soils saturated – only the pretreatment of the soils, at different matric potentials, varied. At the time of determination, the water contents of the soils were therefore the same so that no effect of diffusion could be distinguished. Pentland (1967) found that *Coniophora puteana* was unable to grow through a soil at 20 to 25% saturation or wetter unless the soil was sterile. In the absence of any measurement of water potential it is difficult to analyse the data adequately but

Pentland considered that the diffusion of inhibitors in the wetter soils was important. Certainly her statement that 'microhabitats may be more variable in a drier soil than in a wet one because there is less movement of water-soluble products in the drier soil', is important and true for both stimulatory and inhibitory chemicals.

### 5.7.2. Influence of bulk density

Bulk density may affect microbial activity through change in a number of correlated factors. An increase in bulk density at constant gravimetric water content causes an increase in volumetric water content and hence in solute diffusion. The effect of such a change on the infection of peas by *Pythium ultimum*, studied by Kerr (1964), has already been noted.

The presence of a hardpan in the *B* horizon or at the base of the ploughed layer can greatly influence the distribution of plant pathogenic fungi. Above such a hardpan or within a soil of high bulk density, impeded drainage and poor aeration occur and such sites have been associated with high incidence of root rots caused by *Phytophthora* spp. (Fulton, Mortimore and Hildebrand, 1961; Braun and Wilcke, 1962; Mircetich and Keil, 1970). *Aphanomyces euteiches* and *Fusarium solani*, on the other hand, have been shown to be restricted to the ploughed layer of reduced bulk density (Burke, 1968, 1969; Burke, Hagedorn and Mitchell, 1970) although here there is an interaction with the ability of the host roots to penetrate into, and survive within, the compacted lower horizons. The growth of *Ophiobolus graminis* along wheat roots is reduced by increasing bulk density (Winter, 1939, 1940). The precise mechanism operating in all these cases, is as yet unknown.

A more general circumstance where change in bulk density occurs is in the soil surrounding roots. As a root penetrates, the soil particles are displaced and the bulk density of the surrounding soil increases. Adjacent to a probe, simulating a root, a thickness of soil comparable to the radius of the probe itself is compressed to a maximum bulk density, with a minimum voids system. A lesser increase in bulk density occurs up to a distance of about 3·5 times the root radius from the root surface (Farrell and Greacen, 1966; Barley and Greacen, 1967). The implications of this increased bulk density for rhizosphere organisms has not yet been investigated.

# 6 The influence of oxygen

The literature records many experiments in which change in soil water content evokes a biological response. It often appears that the response is not to a change in any of the factors considered in the previous sections but to some change, associated with soil water content, in the gaseous environment. Some previous work

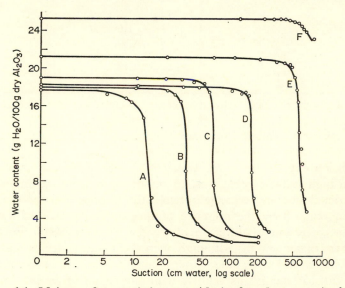

Fig. 6.1. *Moisture characteristic curves (drying boundary curves) of six aluminium oxide grits.* (After Griffin, 1963*b*)

with a species of *Curvularia* (Griffin, 1963*b*) provides a convenient example.

Systems similar to that of Fig. 5.6 were used to control the matric potential of water held in a series of sterile aluminium oxide grits, the moisture characteristics of which are shown in Fig. 6.1. Within the grits were buried small pieces of an agar inoculum of the fungus, each piece placed at the centre of a cross formed of two sterile bundles of hair that acted as a nutrient substrate. Growth of the fungus along the hair was measured at 72 hr (Table 6.1). Considerably greater growth occurred in the systems in which many of the voids were air-filled than in water-saturated systems. Within the range of potentials used (−0·001 to −0·58 bar), matric potential *per se* did not affect growth nor could antagonism or solute

Table 6.1. *Diameter of colonies (mm) of* Curvularia *sp. in grits of various pore sizes and at different suctions* \* *(after Griffin, 1963b)*

|  | Suction (cm water) | | | | | |
|---|---|---|---|---|---|---|
|  | 1 | 22 | 40 | 100 | 200 | 594 |
| Grit A | 4† | *12* | *16* | *17* | *19* | *17* |
| B | 4 | 6 | *11* | *13* | *15* | — |
| C | 4 | 4 | 5 | *14* | *11* | *16* |
| D | 8 | 6 | 7 | 8 | *17* | — |
| E | 8 | 4 | 5 | 8 | 6 | *12* |
| F | 8 | 4 | 8 | 7 | 5 | 4 |

\* Italic type indicates that the pores of the grit were air-filled at the relevant suction.
† Diameter of original inoculum.

diffusion explain the data. Restriction of growth in the systems with high water content was, therefore, tentatively attributed to restricted gaseous exchange between the hyphae and the atmosphere, caused by the presence of an intervening layer of water. (The coefficient of diffusion of oxygen in air is 10,000 times as great as that in water.) Although such a conclusion seems reasonable in this and many other experiments, it has proved exceedingly difficult to establish its validity with any rigour or completeness. To facilitate analysis here, the gases in the soil are divided into three categories – oxygen, carbon dioxide and other volatile products of microbiological action.

### 6.1. Oxygen in the soil

Although some fungi are able to make sparse growth anaerobically (Curtis, 1969) and to make some use of chemicals as alternative electron acceptors to oxygen (Gunner and Alexander, 1964), it is doubtful if the activity of such fungi in soil is ever significant in the absence of oxygen. Active fungi and aerobic bacteria, therefore, act as sinks for oxygen within the soil. Most of the oxygen is taken up through a cytochrome oxidase and is linked with respiration and oxidative assimilation (Beevers, 1961). Smaller quantities, however, are utilized by other terminal oxidases, such as the various phenol oxidases.

It is conventional, when thinking of the biological role of oxygen, to think largely in terms of concentration (or partial pressure) and its effects on growth or some other activity. Such a simple approach, when applied to soil, leaves out of account other factors of equal weight, particularly those associated with oxygen diffusion. Elsewhere (Griffin, 1968a), I have reviewed in greater detail many factors relevant to the biological role of oxygen in soil and have shown that the key to understanding lies in the interrelationship between the uptake, concentration and diffusion of oxygen. Under steady state conditions, the interrelationships are as follows:

(a) The rate of oxygen uptake by a structure is exactly balanced by the rate of oxygen diffusion to that structure.

(b) The rate of oxygen diffusion to the surface of the structure will be determined by three factors. (i) The difference in concentration between that at the surface of the structure and that in the water at the interface between the water and gas-filled pore. (ii) The difference in concentration between that in the pore adjacent to the water surface and that in the atmosphere external to the soil. (iii) The geometry of the system. This last factor includes both the distance and the shape of the pathways by which oxygen reaches all parts of the surface of the structure from the atmosphere. Even in simple systems, factors introduced by geometry are not easy to treat mathematically and it is necessary to make rough generalizations and approximations when the system is as complex as a natural soil. (iv) The diffusion coefficients of the medium around the structure.

(c) It follows from the above that the rate of oxygen diffusion will fall if either the diffusion coefficient or the oxygen concentration at the water/gas interface is lowered. More importantly in soil, the rate will fall if the geometry of the system is altered, for instance by rainfall, so that the effective path for diffusion is lengthened. The original rate of diffusion will be maintained only by a compensating reduction in oxygen concentration at the surface of the structure.

(d) The oxygen concentration at the surface of the structure cannot, however, fall below a certain critical value, characteristic of the structure, without causing a reduction in the rate of oxygen uptake. This is because the rate of oxygen uptake by a terminal oxidase is sensitive to oxygen concentration being governed by the law of mass action, and probably by the Michaelis–Menten relationship:

$$q_m = qc/(K_m + c) \tag{9}$$

where $q_m$ is the measured rate of oxygen uptake at oxygen concentration $c$; $q$, the maximum rate of oxygen uptake; $K_m$, the Michaelis constant of the oxidase, i.e. oxygen concentration at which $q_m = q/2$. The critical concentration at the surface of the structure depends not only upon the affinity of the various terminal oxidases for oxygen but also on the importance and velocity of diffusion between the surface of the structure and the sites of the oxidases.

Initially, responses associated with cytochrome oxidase, and hence with cellular energetics and respiration, will be considered because cytochrome oxidase is the most important oxidase in terms of both volume of oxygen utilized and affinity for oxygen. Its Michaelis constant is of the order of $2.5 \times 10^{-8}$ molar, ten thousand times less than the concentration of oxygen in water in equilibrium with air. (As with water potential, different units may cause confusion. In the atmosphere, oxygen is present to the extent of 21% by volume, or at a partial pressure of $0.21$ atm. The equilibrium concentration in water is then $2.57 \times 10^{-4}$ molar or $8.23 \times 10^{-6}$ g cm$^{-3}$.) The apparent Michaelis constant for oxygen uptake by intact cells of fungi and bacteria (that is, the concentration when the rate of uptake of the whole cell is half-maximal) is higher, but is still less than $3 \times 10^{-6}$ molar (Greenwood, 1968a; Griffin, 1968a). This, however, is not a true Michaelis constant but is determined, in part, by diffusion through the cell wall, the cytoplasm (Johnson, 1967; Griffin, 1968a) and even extracellular mucilage (Nair, White, Griffin and Blair, 1969).

Oxygen is unlikely to have a limiting role on biological activity in soils at low matric potentials because then most voids are filled with gas, in which the diffusion rate of oxygen is $2 \cdot 1 \times 10^{-1}$ cm$^2$ sec$^{-1}$. At higher potentials, oxygen may become limiting because many of the voids are filled with water in which the diffusion rate is $2 \cdot 6 \times 10^{-5}$ cm$^2$ sec$^{-1}$. In a well structured soil, the individual particles are aggregated into crumbs within which the pore radii are very small. The crumbs will thus tend to remain saturated with water over a wide range of matric potentials. Between the crumbs, pore radii are larger and relatively small decreases in potential will greatly alter the proportion of them which are filled with water or gas. It is therefore convenient to consider soil aeration in two stages – within water saturated crumbs and within intercrumb voids.

To understand the problem adequately, it is necessary to treat it mathematically. Many biologists seem to fear equations and I have therefore relegated the mathematical development of a few of the topics to an appendix at the end of this chapter. It should be understood, however, that little more than simple algebraic manipulation is involved and I hope that many will follow through the analyses developed there, for without them it is difficult to appreciate the situation adequately.

Within a crumb of soil, it may be assumed that microorganisms are fairly evenly distributed so that such an aggregate is a potentially respiring 'structure' in the sense of the previous paragraphs. Because of the high rate of respiration of microorganisms and the slow diffusion of oxygen through water, the centres of water-saturated crumbs are likely to be anaerobic if their radii are greater than about 3 mm (Greenwood, 1961; Greenwood and Goodman, 1967), although the precise figure will vary with temperature, amount of organic matter, texture, etc. Conditions within even small crumbs are therefore likely to be heterogeneous. The outer shell may be the site of nitrification and of growth of fungi and aerobic bacteria while the core is simultaneously the site of denitrification and growth of anaerobic bacteria.

This is the crux of the problem in relating microbial activity within crumbs to oxygen in soil. At the level of the individual hypha or cell, activity will be related both to oxygen concentration at the cell surface and to the rate of diffusion of oxygen to the cell. Both are unknown and, with present techniques, unmeasurable because of the microscopic size of the sites involved. Assuming uniform distribution of microorganisms within crumbs, the best

I

that can be done is to estimate the proportion of the crumb that will be respiring under given conditions of the environment, soil type and aggregate size.

Allison (1968) has argued that fungi and bacteria are too large to penetrate the small pores of stable aggregates: if they are there it is by virtue of their incorporation at the time of aggregate formation. If this be true, as seems probable, their activity is unlikely to be long continued because of limits imposed by substrate availability (see p. 104). The more important sites for microbial activity would then be on the various surfaces that form the boundaries of intercrumb voids.

Oxygen diffusion in the pores lying between the crumbs is greatly affected by the proportion filled with water. For a dry material, composed of discrete solid particles:

$$dq/dt = -0{\cdot}66D_oS \, dc/dx \qquad (10)$$

where $dq/dt$ is the rate of gaseous diffusion; $D_o$, the coefficient of diffusion of oxygen in air; $S$, total porosity; $dc/dx$, the concentration gradient and $0{\cdot}66$ is an empirical tortuosity constant (Penman, 1940). Water saturated crumbs of soil are analogous to discrete particles because of the slow diffusion through them. Equation (10) can, therefore, be applied to soils in which the crumbs are water saturated and the intercrumb pores gas filled, so long as $S$ is replaced by $S_v$, the pore space between the crumbs. In most moist soils, water will lie in some of the intercrumb pores as well as within the crumbs. Such water further reduces the gas-filled porosity of the soil, but more importantly, also disrupts the continuity of the gas phase. In such a system with uniform distribution of intercrumb water:

$$dq/dt = -0{\cdot}66D_oS_v(S_g/S_v)^4 \, dc/dx \qquad (11)$$

where $S_g$ is the gas-filled porosity (Currie, 1961$b$). As the rate of diffusion of oxygen depends upon the fourth power of $S_g/S_v$, slight changes in matric potential and hence in the proportion of gas filled, intercrumb voids would be expected to have dramatic effects. In fact, intercrumb water is rarely distributed uniformly and Equation (11) does not then adequately describe the system (Greenwood and Goodman, 1967; Bakker and Hidding, 1970). The general conclusion, however, that rates of diffusion, and therefore of utilization of oxygen, will be sensitive to changes in gas filled porosity is supported by many experimental data.

Greenwood and Goodman (1965) obtained data using columns of soil crumbs, 0·42 to 0·59 mm diam, in which the water content of the intercrumb spaces varied from experiment to experiment whereas the crumbs were always saturated. Overall respiration of the columns was clearly dependent upon gas filled porosity. When a column of soil, originally saturated, was allowed to lose water by evaporation from the top (Yamaguchi, Flocker and Howard, 1967), gas concentrations were markedly affected by both soil water content and depth (and hence by the length of the diffusion path) (Table 6.2). Rixon and Bridge (1968) established for a num-

Table 6.2. *Composition of soil atmosphere at various depths and soil water contents, at 25°C and 38 days after evaporation commenced from the surface of the saturated soil (after Yamaguchi, Flocker and Howard, 1967)*

| Depth (cm) | Water content (% by volume) | Composition (partial pressure, atm) | | |
|---|---|---|---|---|
| | | Oxygen | Carbon dioxide | Nitrogen |
| 5 | 20 | 0·20 | 0·01 | 0·79 |
| 35 | 24 | 0·17 | 0·04 | 0·79 |
| 65 | 27 | 0·04 | 0·13 | 0·83 |

ber of soils that the respiratory quotient fell from greater than two to unity, indicating the establishment of predominantly aerobic respiration, when the gas filled pore space changed from 10 to 20%.

Earlier work on soil aeration has been reviewed by Russell (1961) and Vilain (1963). In general, the composition of the soil atmosphere has been shown to be such that the partial pressure of nitrogen is relatively constant at about 0·79 atm whereas the combined partial pressure of oxygen and carbon dioxide is about 0·21 atm. Most measurements reveal oxygen partial pressures in excess of 0·18 atm but partial pressures as low as 0·1 atm are not uncommon in wet soils. Hack (1956) indicated that the oxygen concentration is lower, and the carbon dioxide concentration is higher, in small pores as compared with large. Within water filled pores the concentration gradient of oxygen is likely to be great so that the concentration will be negligible more than a few milli-metres from an interface.

## 6.2. Oxygen concentration and the growth of soil fungi

In an earlier review (Griffin, 1963a), I noted that soil fungi were relatively insensitive to even severe reductions in oxygen con-centration. Subsequent experiments, some aided by a simple and relatively inexpensive technique for obtaining many different gas

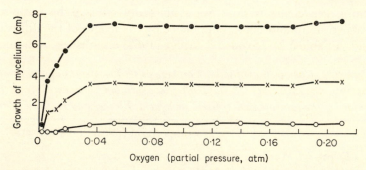

Fig. 6.2. *Relation between growth of* Sclerotium rolfsii *and partial pressure of oxygen.* ○ – 2 days; × – 4 days; ● – 6 days. (After Griffin and Nair, 1968)

concentrations (Griffin, Nair, Baxter and Smiles, 1967), have confirmed this conclusion when germination or linear growth rate are assessed (Wood-Baker, 1955; Waid, 1962; Dukes and Apple, 1965; Brown and Kennedy, 1966; Follstad, 1966; Lockhart, 1967a; Griffin and Nair, 1968; Macauley and Griffin, 1969a). Most of these studies reveal little change in linear growth rate until oxygen concentration is lowered to about 4% in the gas phase (Fig. 6.2). Below this value there is fairly marked difference between species, but nearly all linear growth rates are declining at 1% oxygen. *Phytophthora cinnamomi* and *P. citrophthora* are exceptions, because a reduction in oxygen partial pressure from 0·21 atm to 0·002 atm caused no significant reduction in linear growth rate (K. J. Moore, personal communication). If, however, the effect of oxygen concentration on the dry weight of mycelia grown in agitated liquid culture is measured, fungi appear to be more responsive (Fellows, 1928; Klotz, Stolzy and DeWolfe, 1963; Klotz, Stolzy, DeWolfe and Szuszkiewicz, 1965; Jensen,

1967; Lockhart, 1967b, 1968; Covey, 1970; Wells and Uota, 1970). Such studies show a decline in dry weight with each reduction in concentration of oxygen below 21% in the gas phase, the relationship being almost linear in some cases.

Although factual confirmation is lacking, I suspect that this sensitivity in liquid culture is associated with the problems of diffusion within the fungal pellets noted earlier and discussed more fully in the Appendix to this Chapter. The critical size of the pellets is small even in $2.57 \times 10^{-4}$ molar (21%) oxygen and rapidly decreases with reducing external oxygen concentrations. If the critical size appropriate to each external concentration is exceeded, an anaerobic core will develop, probably with a reduction in translocation and the production of toxic, diffusible metabolites. The likelihood of this occurring will increase with decreasing oxygen concentration in the solution. I believe, therefore, that linear growth rate more accurately measures the direct response of the fungus to changes in oxygen concentration. Furthermore, the diffuse surface growth of most fungi on agar will more nearly approximate the growth on soil substrata than will the dense mycelial development within a pellet.

That growth rate should commence to be affected when oxygen concentration is reduced to about $5 \times 10^{-4}$ molar (4%) is not unexpected. If the response to oxygen has the form shown by other organisms, then 5% reduction in respiration, and hence growth, would be expected when the oxygen concentration is reduced to about twenty times the apparent Michaelis constant, that is at about $6.1 \times 10^{-4}$ molar (5%) oxygen.

If these arguments are accepted, the growth of fungi is unlikely to be affected by changes in oxygen concentration within gas filled pores in continuity with the external atmosphere. Within water filled pores, or gas filled pores not in continuity with the atmosphere, differential tolerance of concentrations lower than $5 \times 10^{-4}$ molar (4%) oxygen may well influence the competitive ability of species. The abundance of *Phytophthora parasitica* compared with the rarity of *Thielaviopsis basicola* in soil at depths exceeding 30 cm has been interpreted in this way (Klotz, Stolzy, DeWolfe and Szuszkiewicz, 1965).

6.2.1. *Oxygen, melanization and the growth of rhizomorphs of* Armillariella elegans.

*Armillariella elegans* and *A. mellea* produce rhizomorphs of apparently identical morphology and physiology. The relationship between oxygen and the growth of rhizomorphs of the former species has been the subject of a comprehensive study by Smith and Griffin (1971) and a selection of the conclusions provides a convenient demonstration of the importance of physical factors in the soil.

The rhizomorphs of *Armillariella* spp. consist of a complex meristematic apex from which is produced a hollow tube, the walls of which consist of closely-packed hyphal cells. At the origin of the rhizomorph, the central canal is continuous with the atmosphere and de Bary was the first to suggest the importance of this gas filled pathway for the rapid diffusion of oxygen to the dense, meristematic apex when the rhizomorph is penetrating a saturated medium. In such a system, the rate of diffusion, and hence total respiration of the apex, will depend in large part on the concentration gradient within the rhizomorph. It might, therefore, be expected that the rate of growth of the rhizomorph into columns of agar would be sensitive to changes in the partial pressure of oxygen at its origin over a certain range. This expectation is supported by the data. A mainly theoretical analysis of oxygen diffusion and growth, too long to be reproduced here, largely accounted for the shape of the curves relating the length of rhizomorph at a given age to the partial pressure of oxygen at the origin of the rhizomorph (Fig. 6.3). As with the other systems previously considered in this chapter, the analysis was based upon the effect of oxygen on cytochrome oxidase. At least one other terminal oxidase is known to be active in *A. elegans* and to this attention will now be turned.

The apex of a rhizomorph growing in a water saturated environment remains white, but the exterior of the older parts of the rhizomorph become dark brown, or 'melanized'. Should the apex emerge into the air for more than a short time, it rapidly becomes brown, because of an intercellular deposit of melanin in the rind, and growth stops at this time. True melanin is produced by the action of tyrosinase on tyrosine but the terms 'melanin' or 'melanization' are frequently used in a looser sense in connection with the action of a number of polyphenol oxidases on substituted phenols. Such a polyphenol oxidase, laccase, is possessed by *A. elegans* and, being a terminal oxidase, its rate of reaction is affected by oxygen concentration. Indeed, it is markedly affected by oxygen concentration throughout the range 0 to 20% for its Michaelis

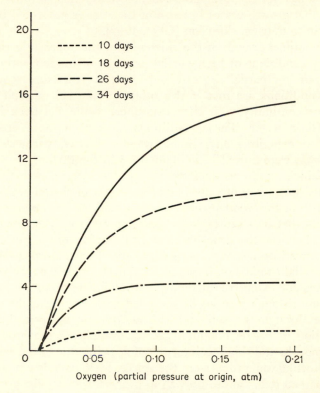

Fig. 6.3. *Influence of oxygen on the lengths of rhizomorphs of* Armillariella elegans *after various periods of growth. All rhizomorphs were growing into columns of nutrient agar but were exposed at their origins to atmospheres containing different partial pressures of oxygen.* (After Smith and Griffin, 1971)

constant was shown to be about 4·22% (5·19 × 10⁻⁵ molar) oxygen. Beneath a water film or in a saturated environment, the concentration of oxygen adjacent to the apex is reduced and the rate of melanization is at least halved compared with the rate in contact with the atmosphere. The data are thus in accord with the hypothesis that rhizomorph elongation ceases when the apex is not covered by a thick film of water because the concentration of oxygen then becomes sufficiently high to induce rapid melanization. A causal relationship between melanization and cessation of growth has yet to be demonstrated. Should this be done, it is likely that growth ceases because of the barrier presented by the melanin to nutrient diffusion (Chet, 1969).

For optimal growth of the rhizomorph, the apparently contradictory conditions of high partial pressure of oxygen within the apex but low partial pressure outside it are therefore required. The conditions are met if the origin of the rhizomorph is in gaseous continuity with the atmosphere while the apex is enveloped in water. These conditions have been more precisely specified experimentally. Small wooden blocks, permeated by *A. elegans* were buried in soil contained in the Büchner funnel of the Haine's apparatus depicted in Fig. 5.6. Matric potential was adjusted to a series of values and the weight of rhizomorphs produced from the wooden blocks were found. It can be seen from Fig. 6.4 that in a saturated or near saturated soil, gaseous continuity between the atmosphere and the surface of the buried noculum was insufficient to permit growth of rhizomorphs. With a further slight reduction in potential, however, more pores became gas filled and extensive rhizomorph development occurred, oxygen diffusing through a series of gas filled pathways from the atmosphere through the soil and rhizomorph to its apex. Further reduction in matric potential greatly diminished the growth of rhizomorphs and this was interpreted as being due to the removal of protecting water films from the apices. Field data do not suggest any marked reduction in growth in near-saturated soils, probably because inocula there are in gaseous connection with the atmosphere through the tree stump. Furthermore, the rhizomorphs when crossing gas filled voids produce abortive side branches which act as 'breathing holes' and thus allow oxygen to enter the rhizomorph at points along the length of the rhizomorph. Field data do, however, indicate a reduction in activity of this fungus in dry soil, in accordance with the laboratory experiment.

Dark fungal pigments ('melanins') have recently been implicated in the resistance of hyphae and conidia to lysis by enzymes such as proteases, chitinase and glucanase (Lockwood, 1960; Potgeiter and Alexander, 1966; Bloomfield and Alexander, 1967; Jones and

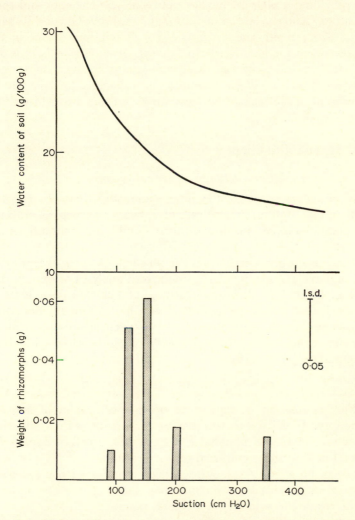

Fig. 6.4. *Moisture characteristic curve* (*drying boundary curve*) *of a soil in which blocks of wood, previously colonized by* Armillariella elegans, *were buried. The histograms depict the weight of rhizomorphs produced from the buried blocks at various matric suctions* (*see text*). (After Smith and Griffin, 1971)

Webley, 1968) and it seems probable that they act mechanically by providing a physical barrier between lytic enzymes and their substrate (Kuo and Alexander, 1967; Chet, 1969; Old and Robertson, 1970b; Smith and Griffin, 1971). Their importance in the protection of the long-lived rhizomorph of *A. elegans* is thus apparent. In a wider context, the interrelationships between oxygen concentration, rate of melanization and resistance to lysis may be significant in the ecology of many fungi but this is a field waiting to be explored.

## 6.3. Appendix to Chapter 6

### 6.3.1. *Oxygen diffusion within a spherical system*

The problems posed by the interrelationships between concentration, rate of diffusion and rate of uptake of oxygen, as outlined on pp. 115-116, are insoluble unless simplifying assumptions are made.

For reasons given earlier, all the terminal oxidases except cytochrome oxidase may be disregarded from analyses of oxygen diffusion. Next, the Michaelis constant of cytochrome oxidase is so small that it may be assumed, as a first approximation, that uptake of oxygen is maximal if any oxygen is present at all at the site of the oxidase. More formally, using the same symbols as in Equation (9) (p. 116):

$$q_m = q \text{ if } c > 0; q_m = 0 \text{ if } c = 0 \tag{12}$$

Further analysis is dependent upon specification of a given geometry. It will be instructive to commence with a spherical structure, such as a moulded soil aggregate, a ball of hyphae (as often found in shake cultures) or a sclerotium.

It can be shown (Rashevsky, 1960) that for such a spherical system, under steady state conditions and with the assumptions made relating to cytochrome oxidase, that:

$$D\left(\frac{d^2c}{dr^2} + \frac{2dc}{r\,dr}\right) = q \text{ (for } c > 0) \tag{13}$$

or, in alternative form:

$$D\frac{d}{dr}^2\left(r^2\frac{dc}{dr}\right) = qr^2 \tag{14}$$

where $D$ is the diffusion coefficient of oxygen with the sphere ($cm^2\ sec^{-1}$); $r$, the radius within the sphere (cm); and $c$, the concentration of oxygen at the site of cytochrome oxidase ($g\ cm^{-3}$).

Upon integration, Equation (14) becomes

$$dc/dr = qr/3D + A/r^2 \qquad (15)$$

and:

$$c = qr^2/6D - A/r + B \qquad (16)$$

where $A$ and $B$, the constants of integration, will depend upon the particular boundary conditions.

Let us now set boundary conditions such that the radius of the surface of the sphere is $R$, the concentration of oxygen at the surface of the sphere is $c_R$ whereas that at the centre is $c_o$. Formally, these boundary conditions are:

$$c = c_R \text{ at } r = R; \text{ and } dc/dr = 0 \text{ and } c = c_o \text{ at } r = 0 \qquad (17)$$

By substituting these boundary conditions into Equations (15) and (16), it can be shown that $A = 0$ and $B = c_o$ and hence that:

$$c_R = c_o + qR^2/6D \qquad (18)$$

By slightly modifying the boundary conditions, an important equation can be derived. Assume that the concentration of oxygen at the centre of the sphere (and only there) is zero ($c_o = 0$). The whole sphere will now be *just* utilizing oxygen at its maximal rate: any reduction in oxygen concentration at the surface of the sphere will lead to the development of an anaerobic core and a decrease in the over-all rate of uptake by the sphere. Let us denote this critical concentration at the surface as $c'_R$. Then, with $c_o = 0$, equation (18) becomes:

$$c'_R = qR^2/6D \qquad (19)$$

Alternatively, if $c_R$ is considered to be constant and equal to the concentration of oxygen in water in equilibrium with the atmosphere, then there will be a corresponding critical radius ($R'$) which *just* permits maximal respiration throughout the sphere, that is:

$$c_R = qR'^2/6D \qquad (20)$$

Clearly, the larger the rate of oxygen uptake or the smaller the diffusion coefficient, the smaller must be the radius of the sphere if an anaerobic core is not to develop. In experiments using fungal pellets in shake culture, it has been possible to prove the validity of Equation (20), and hence of the approximations embodied in Equation (13). The rate of respiration of fungi is rapid and the critical radii for the pellets correspondingly small. Approximate values for $R'$ are 0·1 mm for *Aspergillus niger* (Yano, Kodama and Yamada, 1961), 0·2 mm for *Myrothecium verrucaria* (Darby and Goddard, 1950) and 0·4 mm for *Penicillium chrysogenum* (Phillips, 1966). The peripheral hyphae of submerged, spherical colonies remain in an aerobic environment and therefore continue to grow if adequate nutrients are present in solution (Trinci, 1969). Hyphal masses around small organic particles in saturated zones of a soil, however, are likely to remain small because translocation is inhibited within an anaerobic core.

The oxygen relations of spherical soil aggregates have been extensively analysed (Currie, 1961a; Greenwood, 1961, 1962, 1963; Greenwood and Berry, 1962; Greenwood and Goodman, 1967) and some of the more important conclusions have been noted in Chapter 6. Their derivation is assisted by a further development of the analysis which permits a more quantitative understanding of the situation.

Assume now that oxygen diffuses into the sphere only as far as $r = r_b$. (The central portion of radius $r_b$ is, therefore, anaerobic.) The boundary conditions are thus:

$$c = 0 \text{ and } dc/dr = 0 \text{ when } r = r_b \qquad (21)$$

From Equations (15), (16) and (21)

$$A = -qr_b{}^3/3D \text{ and } B = -qr_b{}^2/2D$$

Hence:

$$c = qr^2/6D + qr_b{}^3/3Dr - qr_b{}^2/2D \qquad (22)$$

and, considering the relevant concentration at the surface,

$$c_R = qR^2/6D + qr_b{}^3/3DR - qr_b{}^2/2D \qquad (23)$$

Re-arranging,

$$6Dc_R/qR^2 = 1 + 2r_b{}^3/R^3 - 3r_b{}^2/R^2 \qquad (24)$$

But the total rate of oxygen uptake by the respiring portion of the sphere will be measured experimentally as the total, but reduced, rate of oxygen uptake for the whole volume of the sphere. That is:

$$4\pi(R^3 - r_b{}^3)q/3 = 4\pi R^3 q_m/3 \qquad (25)$$

where $q_m$ is the measured rate of oxygen uptake per unit volume of the whole sphere of radius $R$ when $c = 0$ at $r = r_b$

Therefore:

$$q_m/q = 1 - (r_b/R)^3 \qquad (26)$$

and:

$$r_b/R = (1 - q_m/q)^{\frac{1}{3}} \qquad (27)$$

From Equations (24) and (27):

$$6Dc_R/qR^2 = 3 - 3(1 - q_m/q)^{\frac{2}{3}} - 2q_m/q \qquad (28)$$

But, from Equation (20):

$$R'^2 = 6Dc_R/q \qquad (29)$$

so that, from Equations (28) and (29):

$$R'^2/R^2 = 3 - 3(1 - q_m/q)^{\frac{2}{3}} - 2q_m/q \qquad (30)$$

For water-saturated soil spheres in the presence of atmospheric oxygen, it is thus possible to relate relative aerobic volumes and rates of oxygen uptake to radii of spheres. Once again, experimental data are in close accord with expectations based on diffusion analysis and so support the basic premises. In the light of this analysis the composition of the gas phase of a soil is of limited significance in interpreting changes in biological activity, particularly as the gas samples are mainly drawn from the large intercrumb voids. Indeed, such a sampling may give misleading results, for Currie (1961a) has pointed out that if the overall respiration of a volume of soil is reduced 'because the crumbs are partly anaerobic, then the anomalous situation arises in which intercrumb conditions apparently improve even though aeration as a whole deteriorates'.

## 6.3.2. Polarographic technique for oxygen determination

It is clear that if a structure lying within a wet soil is to be able to continue to respire at a given rate, then diffusion of oxygen must balance the uptake. Such a consideration has led to the development of techniques to characterize diffusion and, as they are frequently employed, an understanding of the essential theory is advantageous. If a bare platinum electrode is inserted into a soil and polarized by an E.M.F. of about 0·6 volt, then oxygen is reduced at the surface of the platinum. The electrode thus acts as a sink for oxygen, analogous to a respiring structure. The rate at which oxygen is reduced per unit surface area of electrode and hence the rate of diffusion to the electrode can be calculated from measurement of the resultant electrical current (Lemon and Erickson, 1952, 1955; Poel, 1960; Birkle, Letey, Stolzy and Szuszkiewicz, 1964; Letey and Stolzy, 1964; Stolzy and Letey, 1964a, b; Black and Buchanan, 1966; McIntyre, 1966a, b 1967; van Doren and Erickson, 1966). Normally, the bare platinum is a 4 mm length of 22-gauge wire projecting from the sealed tip of a glass tube containing the necessary electrical connections. Analysis cannot proceed without specification of the geometry of the system and diffusion is assumed to have cylindrical co-ordinates. Thus, the electrode, radius $R$, is assumed to be at the centre of a water-saturated soil cylinder of radius $r_p$ beyond which is a gas phase, differing little in composition from the atmosphere. Oxygen uptake in the soil itself is assumed to be negligible. Such a model is clearly an extreme simplification but it does permit analysis. The equation governing diffusion from the gas phase through the soil to the electrode surface is:

$$D_e\left(\frac{d^2c}{dr^2} + \frac{1}{r}\frac{dc}{dr}\right) = 0 \tag{31}$$

where $D_e$ is the diffusion coefficient of oxygen through the soil cylinder.

Integrating:

$$dc/dr = A/r \tag{32}$$

and:

$$c = A\ln r + B \tag{33}$$

Let:

$$c = c_p \text{ at } r = r_p \text{ and } c = c_R \text{ at } r = R \tag{34}$$

Then:

$$A = (c_p - c_R)/r(\ln r_p/\ln R) \tag{35}$$

Now the rate of arrival of oxygen per unit area of electrode surface at the electrode surface, i.e. the flux, $f$, at $r = R$, is given by Fick's

Law as the product of the diffusion gradient and the diffusion coefficient. That is:

$$f = D_e \, dc/dr \tag{36}$$
$$= D_e(c_p - c_R)/R(\ln r_p/R) \quad \text{(from Equation (35))} \tag{37}$$

Hence:

$$\ln (r_p/R) = D_e(c_p - c_R)/Rf \tag{38}$$

and:

$$r_p - R = R(e^{(c_p - c_R)D_e/Rf} - 1) \tag{39}$$

where $(r_p - R)$ is the effective length of the pathway for oxygen diffusion.

Equation (39) is generally applicable within the limits of the assumptions used. For an electrode, $c_R = 0$ (because oxygen is completely reduced) and the equation reduces to

$$r_p - R = R(e^{c_p D_e/Rf} - 1) \tag{40}$$

Because $R$ is a constant and $c_p$ is assumed to be a constant, $f$ is a measure of the effective length of the diffusion path and is known, as noted above, from readings of current.

Values of $f$ thus characterize the relationship between the electrode and oxygen within the soil. The limiting conditions of the oxygen regime for the activity of various organisms might be expected to be correlated with certain values of $f$. Such an expectation has been the background for many experiments and a large body of literature has developed but any correlation between $f$, so determined, and biological activity must be largely empirical and this is especially so for microorganisms. The basic assumption of cylindrical diffusion may be acceptable in the case of the electrode and roots but is certainly invalid for microorganisms which will normally lie on a solid substrate through which diffusion will often be negligible. Furthermore, the equations depend on $R$, which is many orders of magnitude greater for the electrode than for any hyphae, and relative sizes of the component parts of the system affect the flux in other more complex ways (Smiles and Griffin, 1966). Of greatest significance, however, is the difference in behaviour between the electrode and a microorganism when the diffusion gradient is changed. If the gradient increases, the electrode responds by reducing and hence utilizing oxygen at a greater rate, $c_R$ remaining at zero. With a micro-organism, however, the evidence suggests that the rate of respiration of individual cells is independent of oxygen concentration at least until the concentration falls below 5% and is not halved until a concentration of 0·25%. Indeed, the proved validity of the diffusion analyses noted above indicates that the assumptions made in Equation (12) do little violence to the truth. Hence, if the oxygen concentration gradient outside the already respiring cell of a microorganism increases, the system responds not by increased consumption but by reducing the gradient to its original value by a compensating increase in the concentration at the cell surface. For a microorganism, $c_R$ is rarely zero nor is it a constant.

To summarize, a quantitative change in a given biological activity may be statistically correlated with a certain oxygen diffusion rate derived from the use of the polarographic technique. The magnitude of such a rate is, however, not directly related to that biological activity and is certainly not the rate of oxygen diffusion to unit surface area of any microorganism. As long as this limitation is understood, the polarographic technique can be of considerable value.

# 7 The influence of carbon dioxide and volatile chemicals

## 7.1. Carbon dioxide

### 7.1.1. *Carbon dioxide in the soil atmosphere*

The role of carbon dioxide in soil poses all the problems already met in the consideration of oxygen but also new ones associated with the chemical activity of this gas. It is, however, extremely difficult to quantify or analyse theoretically the diffusion of carbon dioxide in soil because few, if any, of the basic data are available. Macfadyen (1970) has discussed various ways of measuring and controlling carbon dioxide concentration in the field and laboratory. Techniques involving diethanolamine must be used with caution because of the possible evolution of ethylene (Forsyth and Eaves, 1969).

The partial pressure of carbon dioxide in the soil atmosphere usually lies between 0·002 and 0·02 atm (Russell, 1961) but values of 0·1 atm or higher have been recorded at depth or in soil of high water content (Leonard, 1945; Burges and Fenton, 1953; Russell, 1961; Yamaguchi, Flocker and Howard, 1967). Recently, Greenwood and Nye (*fide* Greenwood, 1968*a*) have argued on theoretical grounds that the partial pressure of carbon dioxide around a respiring root will rarely exceed 0·01 atm and such a prediction has been supported by their experimental data. Although partial pressures of carbon dioxide of more than a few hundredths of an atmosphere are therefore unlikely to be encountered in soil, evidence on the effects of up to 0·2 atm will now be considered.

### 7.1.2. *Partial pressure of carbon dioxide and the growth of soil fungi*

Many studies have shown that carbon dioxide, in the presence of 0·21 atm oxygen, has little effect on the rate of linear growth of most fungi until its partial pressure exceeds 0·1 atm (Macauley and Griffin, 1969*a*; Smith and Griffin, 1971; for earlier references, see

K

Griffin, 1963*a*). Even fungi such as *Penicillium nigricans* (Burges and Fenton, 1953) *Sclerotium rolfsii* (Griffin and Nair, 1968) (Fig. 7.1) and *Cochliobobolus sativus* (Macauley and Griffin, 1969*a*)

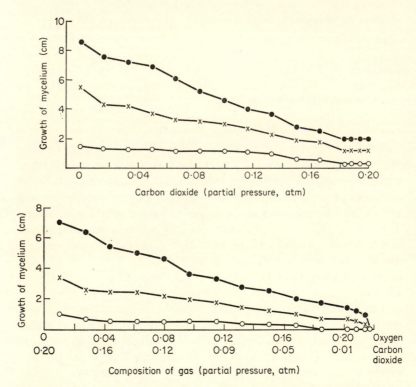

Fig. 7.1. *Relation between growth of* Sclerotium rolfsii *and* (a) *partial pressure of carbon dioxide and* (b) *inverse ratios of partial pressures of oxygen and carbon dioxide* (0·0002 *atm oxygen:* 0·2 *atm carbon dioxide to* 0·21 *atm oxygen:* 0 *atm carbon dioxide*).

○ – 2 days; × – 4 days; ● – 6 days.

(After Griffin and Nair, 1968)

whose growth rates are half-maximal at 0·1 atm carbon dioxide are therefore unlikely to be affected deleteriously within gas filled pores. On the data so far published, only *Sclerotinia minor* (Louvet and Bulit, 1964) is likely to be adversely affected, for its growth rate is halved at 0·04 atm carbon dioxide. Indeed, the contrary may often be the case. Thus, germination of ascospores and formation of perithecia of *Chaetomium globosum* does not occur in the absence of carbon dioxide and is stimulated by partial pressures in excess of atmospheric (Buston, Moss and Tyrrell, 1966). Species of *Fusarium* show a more complicated response because germination of chlamydospores, hyphal growth rate and sporulation are enhanced whereas chlamydospore formation is depressed by increasing carbon dioxide (Stover and Freiberg, 1958; Newcombe, 1960; Bourret, Gold and Snyder, 1965, 1968). The tolerance of hyphal growth of *F. oxysporum* and *F. solani* to very high partial pressures of carbon dioxide has led to their use in selective techniques for isolation from soil (Park, 1961; Bouhot, Bulit and Louvet, 1964).

Durbin (1959) and Burges and Fenton (1953) have correlated tolerance to carbon dioxide with prevalence at depth within the soil but the significance is obscure because of the probable rarity of occurrence of partial pressures of carbon dioxide exceeding 0·05 atm. A similar query must lie against the significance of the marked depression in competitive saprophytic activity of *Rhizoctonia solani* in soils exposed to partial pressure of carbon dioxide exceeding 0·1 atm (Papavizas and Davey, 1962).

As with response to oxygen, response to carbon dioxide is more marked if growth is measured in terms of dry weight of the colony grown in shake culture (Fellows, 1928, Burges and Fenton, 1953; Jensen, 1967). Once again, I suspect that diffusion within the fungal pellet is an interfering factor, this time it being diffusion of carbon dioxide and not oxygen.

### 7.1.3. *Aqueous solutions of carbon dioxide*

The implications of the presence of carbon dioxide in aqueous solutions, particularly those containing calcium or magnesium ions, are profound, but their precise expression lies outside the scope of this book. The interested reader should consult the papers by Ponnamperuma, Martinez and Loy (1966) and Ponnamperuma (1967) and the references therein. The problem in its simplest form is presented by a solution of carbon dioxide in water. The following reactions, somewhat simplified, occur

$$CO_2 \text{ (gas)} + H_2O \rightleftharpoons H_2CO_3$$
$$H_2CO_3 \rightleftharpoons H^+ + HCO_3^-$$
$$HCO_3^- \rightleftharpoons H^+ + CO_3^{2-}$$

The various dissociation constants and their values at 25°C, are given by the following equations.

$$K_{1A} = (H^+)(HCO_3^-)/(H_2CO_3) = 1.72 \times 10^{-4} \qquad (41)$$
$$K_1 = (H^+)(HCO_3^-)/(CO_2)$$
$$= (H^+)(HCO_3^-)/\gamma P_{CO_2} = 4.45 \times 10^{-7} \qquad (42)$$
$$K_2 = (H^+)(CO_3^{2-})/(HCO_3^-) = 4.69 \times 10^{-11} \qquad (43)$$

where $K_{1A}$, $K_1$ and $K_2$ are the first, apparent first and second dissociation constants, parentheses indicate activities, $\gamma$ is Henry's low constant ($0.0344$ mol kg$^{-1}$ atm$^{-1}$ at 25°C) and $P_{CO_2}$ is the partial pressure of carbon dioxide (atm).

From Equation (41), it can be seen that a given activity or concentration of bicarbonate ion can only be maintained if the atmosphere over the solution contains the equilibrium partial pressure of carbon dioxide. Forced passage of air through a solution containing bicarbonate ions will rapidly reduce the concentration to the value in equilibrium with air, an occurrence not apparently realized by all workers. An apparent effect of bicarbonate ion in aerated solutions is more likely to be a response to the cation or to decreased osmotic potential.

If the hydrogen ion concentration of an aqueous system is fixed by a buffering agent and various partial pressures of carbon dioxide are chosen, the resultant activities can be calculated from Equations (41) to (43), as follows:

$$(HCO_3^-) = K_1 \gamma P_{CO_2}/(H^+) \qquad (44)$$
$$(CO_3^{2-}) = K_2(HCO_3^-)/(H^+) \qquad (45)$$
$$(H_2CO_3) = (H^+)(HCO_3^-)/K_{1A} \qquad (46)$$

An inspection of Table 7.1 reveals that, at a given partial pressure of carbon dioxide, an increase of one unit in $p$H increases the activity (or, approximately, concentration) of bicarbonate ion tenfold and that of carbonate ion one hundredfold. Even at high $p$H and high partial pressure of carbon dioxide, the concentration of carbonate ion remains low and can probably be neglected. The concentration of bicarbonate ion, however, is within the range known to affect metabolism in fungi (Cantino, 1966). Present evidence (Griffin, 1965; Macauley and Griffin, 1969b) suggests that fungi respond to bicarbonate ion concentration and not directly to carbon dioxide partial pressure so that no uniform response to carbon dioxide at different hydrogen ion concentrations can be expected. Within soil, the situation is greatly complicated by the presence of many other ions but it now seems unlikely that measurements of partial pressure of carbon dioxide within the soil environment will be of much significance unless $pH$ is measured simultaneously. Data on bicarbonate ion concentration in soil microsites are not available in connection with soil microorganisms.

Table 7.1. *Activities of ions in aqueous solutions of carbon dioxide at various hydrogen ion concentrations*

| $pH$ | Partial pressure $CO_2$ (atm) | $CO_2$ | Activities $H_2CO_3$ | $HCO_3^-$ | $CO_3^{--}$ |
|---|---|---|---|---|---|
| 4 | $3 \times 10^{-4}$* | $1.04 \times 10^{-5}$ | $2.68 \times 10^{-8}$ | $4.95 \times 10^{-8}$ | $2.16 \times 10^{-4}$ |
|   | $1 \times 10^{-1}$ | $3.44 \times 10^{-3}$ | $8.90 \times 10^{-6}$ | $1.53 \times 10^{-5}$ | $7.18 \times 10^{-12}$ |
| 6 | $3 \times 10^{-4}$ | $1.04 \times 10^{-5}$ | $2.68 \times 10^{-8}$ | $4.59 \times 10^{-6}$ | $2.16 \times 10^{-10}$ |
|   | $1 \times 10^{-1}$ | $3.44 \times 10^{-3}$ | $8.90 \times 10^{-6}$ | $1.53 \times 10^{-3}$ | $7.18 \times 10^{-8}$ |
| 8 | $3 \times 10^{-4}$ | $1.04 \times 10^{-5}$ | $2.68 \times 10^{-8}$ | $4.59 \times 10^{-4}$ | $2.16 \times 10^{-6}$ |
|   | $1 \times 10^{-1}$ | $3.44 \times 10^{-3}$ | $8.90 \times 10^{-6}$ | $1.53 \times 10^{-1}$ | $7.18 \times 10^{-4}$ |

* Atmospheric concentration.

Although a given partial pressure of carbon dioxide is likely to have an enhanced effect at higher values of $pH$, partial pressures in soil decline with increasing $pH$. It can be shown from Equations (44) to (46) that, if at $pH$ 4 one unit volume of carbon dioxide is required to raise the partial pressure by 0·1 atm, then the approximate number of unit volumes required to produce the same rise at $pH$ 5, 6, 7, 8 and 9 are 1, 1·5, 5·3, 45 and 460, respectively. If the respiration rate of soils is relatively independent of $pH$, then partial pressures of carbon dioxide will always decline with increasing $pH$.

### 7.1.4. *Carbon dioxide : oxygen ratios*

Within soil, enhanced partial pressures of carbon dioxide are always associated with reduced partial pressure of oxygen, often in such a way that the combined partial pressure of the two gases is approximately 0·21 atm. Greenwood (1968b), however, has shown theoretically and experimentally that the decrease in oxygen partial pressure within water-logged soil crumbs was greater than twenty times the increase in carbon dioxide partial pressure. Whatever the precise values may be, it is of interest to study the effects of simultaneous changes in partial pressures of oxygen and carbon dioxide.

The effect of such simultaneous inverse changes on hyphal elongation in *Sclerotium rolfsii* was the same as when carbon dioxide alone was changed (Griffin and Nair, 1968) (Fig. 7.1). Enhanced growth in gas mixtures containing more carbon dioxide and less oxygen then the standard atmosphere may not be unusual. Thus the growth of the rhizomorphs of *Armillariella elegans* was most rapid when the composition of the atmosphere at their origins was about 0·13 atm oxygen and 0·08 atm carbon dioxide (Smith and Griffin, 1971). Such a response, and others reported in the literature (Lockhart, 1967a, b, 1968; Toler, Dukes and Jenkins, 1966), could not have been anticipated from the response to changes in partial pressures of oxygen or carbon dioxide, applied separately, and an interaction is indicated.

The colonization of substrates in soil under atmospheres that vary in composition but in which the combined partial pressures of carbon dioxide and oxygen is 0·21 atm, have been studied (Griffin 1966a – the introduction is largely erroneous; Macauley and Griffin, 1969a). The data do not reveal significant changes in the composition of the colonizing flora until the carbon dioxide: oxygen ratio exceeds unity. Interpretation of the data on the myco-flora at ratios exceeding unity is not easy, but change in partial pressure of carbon dioxide rather than oxygen appears to have been the determining factor. The linear growth rate of *Cochliobolus sativus*, *C. spicifer*, *Chaetomium sp.* and *Curvularia* sp. declines in parallel with their frequency of isolation from the substrate as carbon dioxide increased. On the other hand, increasing frequency of *Trichoderma* sp. and *Gibberella zeae* on the substrate with in-creasing carbon dioxide was not associated with a similar growth response in pure culture. The effect was attributed to increased tolerance to carbon dioxide on the part of these species relative to that of their antagonists.

## 7.2. Volatile chemicals within the soil

There is a rapidly increasing interest in biologically active volatile chemicals within the soil. Many are associated with decomposing plant residues and those characterized are primarily aldehydes and alcohols although ammonia has also been implicated (Gilpatrick, 1969; Owens, Gilbert, Griebel and Menzies, 1969). These products affect fungi and other microorganisms in a number of ways (Latham and Watson, 1967). Most, when in low concentration, increase microbial respiration and numbers but at higher concentrations they become inhibitory (Menzies and Gilbert, 1967; Gilbert, Menzies and Griebel, 1969; Linderman, 1970). In the case of *Verticillium dahliae* (Gilbert and Griebel, 1969) and *Sclerotium rolfsii* (Linderman and Gilbert, 1969), the final effect is deleterious and is associated with increased antagonism after exposure of the soil to the volatile chemicals. Not all such chemicals are associated with the decomposition of relatively large quantities of plant residues. Balis and Kouyeas (1968) provided evidence for the existence in soil of volatile inhibitors to spore germination and suggested their implication in soil fungistasis. In pure culture, nonanoic acid is produced by fungi and this is a stable but volatile inhibitor of germination (Garrett and Robinson, 1969). Volatile sulphides are also implicated in the germination of the sclerotia of *Sclerotium cepivorum* (see p. 64). Finally, sporophore formation in *Agaricus bisporus* appears to depend on the action of bacteria, such as *Pseudomonas putida*, that are stimulated by the volatile aromatic products of metabolism of the fungus (Hayes, Randle and Last, 1969).

The diffusion and concentration of volatile chemicals within the soil must be affected by soil water content and so by a range of factors analogous to those considered for oxygen and carbon dioxide, but I know of no work in this potentially important field. It is tempting to suggest, however, that volatile chemicals may be involved in experiments where change in $p$H or forced aeration of a soil produces a biological response, even though it is difficult to believe that concentrations of either oxygen or carbon dioxide were initially inhibitory (Garrett, 1937; Blair, 1943; Abeygunawardena and Wood, 1957).

# 8 Other soil physical factors

## 8.1. Radiation

Radiation is probably not of great significance because the soil must protect most microorganisms from harmful radiation, mainly ultraviolet light. There is, however, a curious predominance of members of the Dematiaceae and Sphaeropsidales in soils exposed to intense insolation, such as those of deserts and sandbars (Brown, 1958; Griffin, 1960; Nicot, 1960; Gochenaur and Backus, 1967) and in soils from an atomic bomb site (Durrell and Shields, 1960). Such predominance may result from the presence of melanin pigments which absorb ultraviolet radiation and so confer some degree of protection on the propagules (Parmeter and Hood, 1962). Of equal importance, however, might be the thick-walled multicellular resting stages (conidia, bulbils, sclerotia) of many of the fungi. This emphasis on the resting stage may be misplaced. Gochenaur and Whittingham (1967) suggested that an unusual ability to grow in an environment that is both physically rigorous and nutritionally impoverished may also determine the fungal population of sandy areas with few higher plants.

## 8.2. Temperature

Of all the physical variables of biological significance, temperature is perhaps the most obvious. Surface temperatures vary with latitude and altitude and change cyclically each day and each year. Soil temperature is also greatly influenced by the presence or absence of vegetation, by water content and by depth within the soil: the basic physical concepts are well presented by Rose (1966). It is sufficient for us here to note that the temperature of surface soils may drop far below freezing point or exceed 60°C and so, at either end of the scale, be unfavourable for fungal growth. Even within one day, a fluctuation of 35°C may occur at the soil surface in temperate zones and be greatly exceeded in deserts (Russell, 1961). With increasing depth diurnal temperature fluctuations are reduced and approach a mean value. The mean value in summer in sub-surface soils in cool, temperate climates may be only 10°C whereas in subtropical and tropical regions the mean may approach 30°C.

In soil, fungi may therefore be exposed to a great range of temperatures and this has led to many experiments to determine the growth response of fungi to various constant temperatures. Illustrative examples are given in Fig. 8.1. These reveal that the

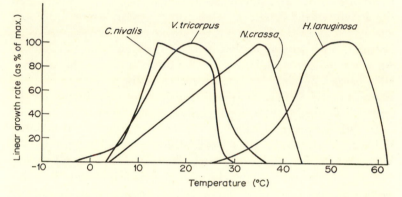

Fig. 8.1. *Influence of temperature on the rate of linear growth of four fungi.* (Based on Anliker *fide* Gaumann (1950) for *Calonectria nivalis*; Isaac (1953) for *Verticillium tricorpus*; Ryan, Beadle and Tatum (1943) for *Neurospora crassa*; Blom & Snider *fide* Emerson (1968) for *Humicola lanuginosa*)

response can vary greatly even though most species have cardinal temperatures of about 5°, 25° and 35°C. Deverall (1965), discussing the ambiguity of the concept of cardinal temperatures has pointed out that the optimum temperature for growth may vary with length of exposure. The optimum for linear extension may also differ from that for antibiotic production. Furthermore, Rishbeth (1968) has shown that the growth rates of rhizomorphs of *Armillaria mellea* on malt agar were 9·8 and 6·8 mm day$^{-1}$ at 25°C (optimum) and 30°C, respectively. Growth in soil, however, was greatest at 22°C and nil at 30°C. Rhizomorphs were not initiated from woody, as opposed to agar, inocula at temperatures exceeding 25°C. Sometimes the differences between optima for different activities is marked, as in *Cercosporella herpotrichoides*, for which

5°C is both the minimum for vegetative growth on agar and the optimum for sporulation on straw (Jørgensen, 1964). Clearly, the concept of cardinal temperatures needs to be applied with care in ecology, because activity and survival depend upon many factors.

Nearly all experiments have been performed at a range of constant temperatures and there is little precise information on the effects of fluctuating temperature regimes *in vitro* similar to those occurring in the field. It appears to have been tacitly assumed that the activity of the fungus is related to the mean temperature in a cycle so long as the temperatures are never so extreme as to be harmful. Such an assumption involves the belief that fluctuations *per se* are neither stimulatory nor inhibitory and that the relationship between growth and temperature is linear throughout the relevant range. In fact, there is little experimental evidence for the first belief and the latter is generally unlikely to be valid.

In soil, temperature usually rises more rapidly than it falls but the cycle is approximately sinusoidal. Burgess and Griffin (1968*b*) have attempted to analyse precisely the response of fungal growth to such a cycle and they developed a mathematical analysis which permitted growth rates under a cyclical regime to be predicted. To test the validity of the analysis, three fungi were grown under sixteen diurnal cycles with median temperatures between 15° and 30°C and amplitudes between 2 and 8°C. Measured and predicted growths were, with few exceptions, in close agreement and gave no support to any assumption that temperature fluctuation *per se* affects the rate of linear extension. The exceptions occurred when the maximum temperature within a cycle exceeded or approached the maximum temperature for growth of the fungus. Then the high temperature caused a deleterious effect for some time after it was reduced. An effect of temperature fluctuation *per se* on growth is suggested in other work (Smith, 1964; Jensen, 1969) but requires confirmation by a more adequate analysis. Clearly, however, data for many other species *in vitro* are required, before an attempt can be made to interpret the role of different diurnal cycles in fungal ecology.

Before considering the effect of temperature on mesothermic fungi in soil, we shall consider fungi that are active at the two extremes of the temperature range, the psychrophiles and thermophiles. Psychrophilic fungi are those able to grow at 0°C and with an optimum at or below 20°C. *Sclerotinia borealis*, which provides an extreme example (Fig. 8.2.), is one of a small group of psychrophilic fungi, including *Typhula idahoensis* and *Calonectria nivalis*, that are able to cause snow-mould by attacking leaves of the Gramineae when buried under snow (Bruehl, Sprague *et al.*, 1966). They are also capable of saprophytic activity under these conditions. Their distribution appears to be determined by their degree of psychrophily, *S. borealis* being restricted to northern regions with exceptionally severe winters whereas *C. nivalis* is common in

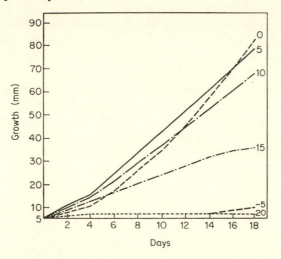

Fig. 8.2. *Influence of temperature on the growth of* Sclerotinia borealis. (After Ward, 1966)

temperate countries. Little is known of the physiological basis of psychrophily although a basis can be suggested for *S. borealis* (Ward, 1966). The optimum temperature for growth of this species was 0°C but that for rate of oxygen uptake was 25°C, which is similar to that of many fungi. It would thus appear that respiration became uncoupled from growth as the temperature rose above freezing-point. The deleterious effects of temperatures above 20°C caused a remarkably long period to ensue before growth was resumed at 0°C, after previous exposure to the higher temperatures.

Thermophilic fungi are those with a maximum temperature for growth at or above 50°C and a minimum at or above 20°C (Cooney and Emerson, 1964; Emerson, 1968). Their occurrence in self-heating composts is not remarkable but their presence in soils, particularly of temperate regions (Apinis, 1962; Eggins and Malik, 1969), was rather unexpected. Their frequency seems too high to be explained by dispersal of spores from other sites and one must conclude that they are, at some periods, active in those soils from which they can be isolated. Even in temperate climates, the temperature of surface soils will permit the growth of thermophilic species on some days, although growth must be intermittent and suboptimal. Microbial thermogenesis in soil is unlikely to produce a rise in temperature of more than a few degrees (Clark, Jackson and Gardner, 1962) and thus is unimportant compared with insolation as a source of heat. The ecology of thermophilic fungi in soil, however, is a promising field for further study.

The respiration of both subantarctic and subtropical soils increased as the temperature increased from 10° to 37°C but then declined from 37° to 50°C (Bunt and Rovira, 1955). Correspondingly, the rate of decomposition of organic matter in soil increases with temperature over much of the field range: wheat straws lost 32·4% of their dry weight in eight weeks at 10°C but 51·7% at 25°C (Burgess and Griffin, 1967). Although these data reflect overall microbial activity, there is no reason to believe that the mesothermic fungal component is radically different from that of the whole in the range from 0° to 30°C. Thereafter, however, a continued increase in respiration is likely to be attributable almost exclusively to bacteria and actinomycetes.

The relationship between temperature response *in vitro* and activity in soil is rarely, if ever, straightforward. The success of *Fusarium oxysporum* and *Cylindrocarpon radicicola* in the competitive saprophytic colonization of dead roots obviously depends on temperature but the precise form of the dependency could not have been foreseen from *in vitro* studies of linear extension (Table 8.1). Indeed, growth rate and colonization were inversely related for *C. radicicola*. Similarly, competitive saprophytic colonization of sterile wheat straw by *Gibberella zeae*, *Fusarium culmorum* and *Cochliobolus sativus* increased as the temperature decreased within the range 10 to 30°C and success was entirely unrelated to linear growth rate (Burgess and Griffin, 1967, 1968b). In all these cases, experiments strongly suggested that the level of antagonism was the major factor that determined successful saprophytism. In general, competition is likely to predominate, as witnessed by respiration data, but antibiosis and exploitation are possibly involved, too. For example, hyphae of *Rhizoctonia solani* are reported to be invaded in soil by *Penicillium vermiculatum* and *Trichoderma* spp. at 28°C but rarely at 18°C (Boosalis, 1956).

Table 8.1. *Effect of temperature on growth and competitive saprophytic colonization by* Fusarium oxysporum *and* Cylindrocarpon radicicola *(after Taylor, 1964)*

| | Growth (diam in eight days (mm)) | | | | Root pieces colonized* (%) | | | |
|---|---|---|---|---|---|---|---|---|
| Temperature (°C) | 10 | 18 | 21 | 25 | 10 | 18 | 21 | 25 |
| F. oxysporum | 6 | 59 | 71 | 80 | 24 | 34 | 46 | 38 |
| C. radicicola | 20 | 51 | 55 | 53 | 16 | 4 | 0 | 0 |

* Substrate – killed sterile dwarf bean roots
  Inoculum – Maize-meal sand inoculum (40%)
      mixed with soil (60%)

## 8.3. Hydrogen ion concentration

Rates of endogenous respiration in fungi are little affected by the hydrogen ion concentration of the external medium over the range $p$H 5 to 8. This is in agreement with the meagre evidence suggesting that fungal cells are well buffered, with little variation in internal $p$H despite changes in external $p$H (Cochrane, 1958). Exogenous respiration and growth, however, are affected by changes in external $p$H and are presumably influenced primarily by systems located at the cell surface and by changes in the permeability of membranes. Many such systems sensitive to hydrogen ion concentration are known, but they are not discussed here because their role in ecology is unknown. Hydrogen ion concentration also affects environmental factors such as the solubilities of minerals (Chapman, 1965) and the ionization of salts and acids independently of the presence of organisms. The response of organisms to change in $p$H is thus likely to depend upon the integration of many factors, some predominating over one part of the range and others over other parts. Because of this, the physiological basis of ecological data concerning $p$H is difficult to evaluate and little progress has been made in this direction for soil fungi.

That there is an ecological differentiation between soil fungi based on $p$H was shown in a classical study by Warcup (1951). He examined, using the soil-plate technique, the mycofloras of a range of natural grassland soils at Lakenheath Warren in England. All lie within a few miles of each other, are light textured and derived from a mixture of chalky boulder clay and interglacial sands overlying chalk. They vary, however, from shallow, highly alkaline soils (grassland $A$) to deep highly acid podsols (grassland $E$). The data (Table 8.2) reveal that fungi were more abundant, in terms of both

Table 8.2. *Some characteristics of the mycofloras of five grassland soils (after Warcup, 1951)*

| Soil | A | B | C | D | E |
|------|------|------|------|------|------|
| $p$H | 8·0 | 7·5 | 6·4 | 4·2 | 3·9 |
| Total number of species | 48 | 62 | 74 | 79 | 69 |
| Average number of species per profile | 23 | 33 | 37 | 35 | 25 |
| Number colonies per g of surface soil ($\times$ 1,000; by dilution plate) | 36 | 59 | 74 | 245 | 429 |

species and isolates, in the acid soils and this is typical of other acid soils. Some species, such as *Penicillium nigricans*, were common in all five soils whereas *Humiocla grisea* and *P. rugulosum* were virtually limited to soil $A$, *P. terlikowskii* to soils $D$ and $E$. Brown (1958) examined the mycofloras of both acid and alkaline coastal dunes and her data for many species are in close agreement with Warcup's. Thus both found *Mortierella isabellina*, *Mucor ramannianus* and *Trichoderma viride* to be commonest in acid systems, *Absidia glauca* and *Mortierella alpina* commonest in alkaline ones. That $p$H was truly the determinative factor in these relationships is further supported by the finding that many of the fungi characteristic of these acid soils were also frequently found in acid bogs and swamps in Wisconsin (Christensen and Whittingham, 1965). The only common environmental factors are acidity and base deficiency. The basis for these preferences is unknown and certainly complex because in culture these fungi do not have the same restricted range of preferred $pH$ as they do in soil.

Changing the hydrogen ion concentration of soil often changes the activity of organisms. Thus *Pythium mamillatum* introduced into a woodland soil of $pH$ 3·7 failed to colonize maize meal sand added subsequently, although colonization, and the formation of oospores and zoosporangia, occurred when the same fungus was added to a garden loam of $pH$ 5·4. Neutralization of the woodland soil with lime permitted colonization and sporulation by the fungus (Barton, 1958).

In unbuffered culture media, the growth of microorganisms leads to rapid changes in $pH$ because anions or cations are absorbed preferentially. The direction of the change is largely determined by the composition of the nitrogen source. Although soil is relatively well buffered, microorganisms can change the $pH$, either directly or indirectly. Probably the best documented instance is that of *Sclerotium rolfsii*, for which $pH$ 4·0 to 5·0 is the optimum in culture and $pH$ 1·4 to 2·0 the minimum. The fungus produces large quantities of oxalic acid, both in culture and in infected hypocotyls of bean. As a result, the $pH$ of the host tissue drops from 5·8 to 4·0. The latter value is optimal for the activity of polygalacturonase and three cellulases produced by this fungus. The combined effect of oxalic acid and the enzymes is thus to cause a rapid breakdown of cellulose and pectic materials, even if the latter are in the form of calcium pectate, which is unaffected by the enzymes alone (Bateman and Beer, 1965; Bateman, 1967, 1969). The ability of *S. rolfsii* to form and tolerate high concentrations of hydrogen ions is clearly important in its utilization of many substrates. Although this biochemical mechanism has not been shown to act saprophytically in soil, there is no reason to doubt that it does. Given a suitable substrate, such a fungus is likely to be able to adjust the $pH$ to the optimum, at least in the zone adjacent to the hyphae, and so be largely independent of the general $pH$ of the soil. That such is the case with *S. rolfsii* is indicated by the failure, or variable success, of soil $pH$ adjustment (even to $pH$ 8·1) to control disease caused by the fungus. (Aycock, 1966.)

A less direct method by which hydrogen ion concentration may be changed has been revealed by a study of soil actinomycetes.

Although certain *Streptomyces* spp. are generally unable to grow in conditions more acid than $p$H 5, activity was shown in some soils between $p$H 4 and 5. Closer investigation revealed that the $p$H adjacent to lysing fungal hyphae was considerably higher because of the liberation of ammonia, presumably from the decomposition of the glucosamine constituents of chitin (Williams and Mayfield, 1971). Thus, these *Streptomyces* spp. were able to grow in these microsites although the bulk of the soil was unfavourable to them.

The situation just discussed raises the issue of heterogeneity of distribution of hydrogen ions within the soil. It is perhaps not unexpected that their distribution will vary between microsites of the type just noted. There is, however, a far more fundamental heterogeneity which has been largely ignored by soil microbiologists. It derives from the existence in soil of highly charged clay colloids of enormous total surface area. Most clay particles are negatively charged so that adjacent to their surfaces there is a predominance of cations, largely hydrogen ions, within the soil water (Rose, 1966). The disproportion between concentrations of anions and cations decreases with distance from the clay surface and eventually becomes nil. The zone in which a disproportion exists is called a 'diffuse double layer' and within it, the hydrogen ion concentration changes, being highest immediately adjacent to the charged surface and lowest at and beyond the boundary of the layer. The thickness of this double layer $\left(\frac{1}{k}\right)$ is given by the equation

$$\frac{1}{k} = \frac{1}{KZ^+\sqrt{M}} \tag{47}$$

where $K$ is a constant, $Z^+$ is the valency of cation, $M$ is the molecular concentration of external solution. In a charged colloidal system such as a soil, the hydrogen ion concentration thus varies with distance from surfaces. In a soil with few metallic cations, most of the hydrogen ions may be present within the large double layers and an overall value for $p$H may come to have a most problematical significance. If metallic cations, particularly of high valency, are common, both $Z^+$ and $M$ in Equation (47) will increase and most of the hydrogen ions will be displaced into the solution external to the reduced double layers. The measured $p$H value then regains significance. It is for this reason that values of hydrogen ion concentration in soil should be determined in suspensions in 0·01 molar $CaCl_2$. Examination of hyphae recovered from soil reveals that many are densely coated in clay particles. Such a coating is likely to have marked effects on the hydrogen ion concentration at the hyphal surfaces and to render it different from that in the external solution, but I know of no studies of this.

The concept of hydrogen ion concentration is deceptively simple, when applied to soil and its more esoteric aspects may well be of significance in microbial ecology. Experiments on one such aspect have been reported and they are highly suggestive if not conclusive (Stotzky and Martin, 1963; Stotzky and Rem, 1966, 1967; Stotzky, 1966a, b; Stotzky and Post, 1967). The distribution in soil of the causal agents of Panama wilt of bananas (*Fusarium oxysporum* f. *cubense*) and histoplasmosis of man and animals (*Histoplasma capsulatum*) was studied. In nearly all cases, soils containing the organisms or in which their spread appeared to be rapid were devoid of certain clay minerals. These minerals were characterized as being crystalline, three-layered hydrous aluminosilicates with marked swelling properties even at reduced water potentials. In liquid culture, it has been shown that such clays, approximating to montmorillonites, have little effect on fungal growth. Their presence, however, compared with controls or other clay minerals, greatly enhances the rate of respiration of many soil bacteria. The effect has been traced to the buffering action of the clays on the *p*H of the media, many bacteria being far more sensitive than fungi to change in *p*H. The efficiency of the montmorillonite fraction, compared with that of the kaolinite or vermiculite fractions, as buffers has been attributed to the combination of high cation exchange capacity and wide lattice expansion even at −15 bar matric potential, expansion being necessary for the free interchange of cations. As a consequence, it has been suggested that the absence, or limited activity, of *F. oxysporum* f. *cubense* and *H. capsulatum* in soils containing montmorillonite clays is caused by an increased bacterial antagonism that is encouraged by the more stable *p*H.

L

# References

Abercrombie, M., Hickman, C. J., and Johnson, M. L. (1954). *A Dictionary of Biology*, 2nd ed. Harmondsworth: Penguin.

Abeygunawardena, D. V. W., and Wood, R. K. S. (1957). Factors affecting germination of sclerotia of *Sclerotium rolfsii*. *Transactions of the British Mycological Society* **40**, 221–31.

Acker, L. (1962). Enzymic reactions in foods of low moisture content. *Advances in Food Research* **11**, 263–330.

Albersheim, P., Jones, T. M., and English, P. D. (1969). Biochemistry of the cell wall in relation to infective processes. *Annual Review of Phytopathology* **7**, 171–94.

Alexander, M. (1961). *Introduction to Soil Microbiology*. London: John Wiley and Sons.

Allison, F. E. (1968). Soil aggregation – some facts and fallacies as seen by a microbiologist. *Soil Science* **106**, 136–43.

Anderson, A. L., and Huber, D. M. (1965). The plate–profile technique for isolating soil fungi and studying their activity in the vicinity of roots. *Phytopathology* **55**, 592–4.

Andrewartha, H. G. (1964). How animals can live in dry places. *Proceedings of the Linnean Society of New South Wales* **89**, 287–94.

Apinis, A. E. (1962). Occurrence of thermophilous microfungi in certain alluvial soils near Nottingham. *Nova Hedwigia* **5**, 57–78.

Apinis, A. E. (1964). Revision of British Gymnoascaceae. *Mycological Papers of the Commonwealth Mycological Institute* **96**, 1–56.

Armolik, N., and Dickson, J. G. (1956). Minimum humidity requirement for germination of conidia of fungi associated with storage of grain. *Phytopathology* **46**, 462–5.

Atkinson, R. G., and Robinson, J. B. (1955). The application of a nutritional grouping method to soil fungi. *Canadian Journal of Botany* **33**, 281–8.

Aycock, R. (1966). Stem rot and other diseases caused by *Sclerotium rolfsii*. *North Carolina Agricultural Experimental Station*, Technical Bulletin **174**, 1–202.

Baker, K. F., Flentje, N. T., Olsen, C. M., and Stretton, H. M. (1967). Effect of antagonists on growth and survival of *Rhizoctonia solani* in soil. *Phytopathology* **57**, 591–7.

Baker, R. (1968). Mechanisms of biological control of soil-borne pathogens. *Annual Review of Phytopathology* **6**, 263–94.

Baker, R. (1970). Use of population studies in research on plant pathogens in soil. In *Root Diseases and Soil-borne Pathogens*, T. A. Toussoun, R. V. Bega and P. E. Nelson, eds., pp. 11–15. Berkeley: University of California Press.

Baker, R., Maurer, C. L., and Maurer, R. A. (1967). Ecology of plant pathogens in soil. VII. Mathematical models and inoculum density. *Phytopathology* **57**, 662–6.

Bakker, J. W., and Hidding, A. P. (1970). The influence of soil structure and air content on gas diffusion in soils. *Netherlands Journal of Agricultural Science* **18**, 37–48.

Balis, C., and Kouyeas, V. (1968). Volatile inhibitors involved in soil mycostasis. *Annales de l'Institut Phytopathologique Benaki* N.S. **8**, 145–9.

Barley, K. P., and Greacen, E. L. (1967). Mechanical resistance as a soil factor influencing the growth of roots and underground shoots. *Advances in Agronomy* **19**, 1–44.

Barnes, M., and Parker, M. S. (1966). The increase in size of mould spores during germination. *Transactions of the British Mycological Society* **49**, 487–94.

Barnes, M., and Parker, M. S. (1967). Use of the Coulter counter to measure osmotic effects on the swelling of mould spores during germination. *Journal of General Microbiology* **49**, 287–92.

Barnett, H. L. (1964). Mycoparasitism. *Mycologia* **56**, 1–19.

Barron, G. L., and Fletcher, J. T. (1970). *Verticillium albo-atrum* and *V. dahliae* as mycoparasites. *Canadian Journal of Botany* **48**, 1137–9.

Barton, R. (1958). Occurrence and establishment of *Pythium* in soils. *Transactions of the British Mycological Society* **41**, 207–22.

Barton, R. (1961). Saprophytic ability of *Pythium mamillatum* in soils. II. Factors restricting *P. mamillatum* to pioneer colonization of substrates. *Transactions of the British Mycological Society* **44**, 105–18.

Bateman, D. F. (1963). Factorial analysis of environment and pathogens in relation to development of the *Poinsettia* root rot complex. *Phytopathology* **53**, 509–16.

Bateman, D. F. (1967). Characteristics of the cellulase system produced by *Sclerotium rolfsii*. *Phytopathology* **57**, 803.

Bateman, D. F. (1969). Some characteristics of the cellulase system produced by *Sclerotium rolfsii* Sacc. *Phytopathology* **59**, 37–42.

Bateman, D. F., and Beer, S. V. (1965). Simultaneous production and synergistic action of oxalic acid and polygalacturonase during pathogenesis by *Sclerotium rolfsii*. *Phytopathology* **55**, 204–11.

Bateman, D. F., and Millar, R. L. (1966). Pectic enzymes in tissue degradation. *Annual Review of Phytopathology* **4**, 119–46.

Bavendamm, W., and Reichelt, H. (1938). Die Abhängigkeit des Wachstums holzzersetzender Pilze von Wassergerhalt des Nährsubstrates. *Archiv für Mikrobiologie* **9**, 486–544.

Beament, J. W. (1965). The active transport of water: evidence, models and mechanisms. In *The State and Movement of Water in Living Organisms*, 19th Symposium of the Society for Experimental Biology, C. E. Fogg, ed., pp. 237–98. Cambridge: Cambridge University Press.

Beevers, H. (1961). *Respiratory Metabolism in Plants*. New York: Harper and Row.

Bhaumik, H. D., and Clark, F. E. (1947). Soil moisture tension and microbiological activity. *Proceedings of the Soil Science Society of America* **12**, 234–8.

Bimpong, C. E., and Clerk, G. C. (1970). Motility and chemotaxis in zoospores of *Phytophthora palmivora* (Butl.) Butl. *Annals of Botany, London* N.S. **34**, 617–24.

Birkle, D. E., Letey, J., Stolzy, L. H., and Szuszkiewicz, T. F. (1964). Measurement of oxygen diffusion rates with the platinum microelectrode. II. Factors influencing the measurement. *Hilgardia* **35**, 555–66.

Black, J. D. F., and Buchanan, A. S. (1966). Polarographic reduction of oxygen at a platinum surface in relation to the measurement of oxygen flux in soils. *Australian Journal of Chemistry* **19**, 2169–74.

Blair, I. D. (1943). Behaviour of the fungus *Rhizoctonia solani* Kühn in the soil. *Annals of Applied Biology* **30**, 118–27.

Bloomfield, B. J., and Alexander, M. (1967). Melanins and resistance of fungi to lysis. *Journal of Bacteriology* **93**, 1276–80.

Blumenthal, H. J. (1965). Carbohydrate metabolism. I. Glycolysis. In *The Fungi*, Vol. 1, G. C. Ainsworth and A. S. Sussman, eds., pp. 229–68. New York: Academic Press.

Bodman, G. B., and Day, P. R. (1943). Freezing points of a group of California soils and their extracted clays. *Soil Science* **55**, 225–46.

Bond, R. D. (1964). The influence of the microflora on the physical properties of soils. II. Field studies on water repellent sands. *Australian Journal of Soil Research* 2, 123–31.

Bond, R. D., and Harris, J. R. (1964). The influence of the microflora on the physical properties of soils. I. The occurrence and significance of microbial filaments and slimes in soils. *Australian Journal of Soil Research* 2, 111–22.

Bonner, J. T. (1948). A study of the temperature and humidity requirements of *Aspergillus niger*. *Mycologia* 40, 728–38.

Boosalis, M. G. (1956). Effect of soil temperature and green-manure amendment of unsterilized soil on parasitism of *Rhizoctonia solani* by *Penicillium vermiculatum* and *Trichoderma* sp. *Phytopathology* 46, 473–8.

Boosalis, M. G. (1964). Hyperparasitism. *Annual Review of Phytopathology* 2, 363–76.

Boosalis, M. G., and Mankau, R. (1965). Parasitism and predation of soil microorganisms. In *Ecology of Soil-borne Plant Pathogens*, K. F. Baker and W. C. Snyder, eds., pp. 374–89. Berkeley: University of California Press.

Borut, S. (1960). An ecological and physiological study on soil fungi of the Northern Negev (Israel). *Bulletin of the Research Council of Israel* 8D, 65–80.

Bouhot, D., Bulit, J., and Louvet, J. (1964). Recherches sur l'écologie des champignons parasites dans le sol. III. Recherche d'une méthode sélective et quantitative d'analyse de *Fusarium oxysporum* f. *melanis* dans le sol. *Annales des Epiphyties* 15, 57–72.

Bourret, J. A., Gold, A. H., and Snyder, W. C. (1965). Inhibitory effect of $CO_2$ on chlamydospore formation in *Fusarium solani* f. *phaseoli*. *Phytopathology* 55, 1052.

Bourret, J. A., Gold, A. H., and Snyder, W. C. (1968). Effect of carbon dioxide on germination of chlamydospores of *Fusarium solani* f. sp. *phaseoli*. *Phytopathology* 58, 710–11.

Bowen, G. D., and Rovira, A. D. (1968). The influence of microorganisms on growth and metabolism of plant roots. In *Root Growth*, W. J. Whittington, ed., pp. 170–99. London: Butterworth.

Braun, H., and Wilcke, D. E. (1962). Untersuchungen über Bodenverdichtungen und ihre Beziehungen zum Auftreten der Kragenfäule (*Phytophthora cactorum*). *Phytopathologische Zeitschrift* 46, 71–86.

Brian, P. W. (1960). Antagonistic and competitive mechanisms limiting survival and activity of fungi in soil. In *Ecology of Soil Fungi*, D.

Parkinson and J. S. Waid, eds., pp. 115–29. Liverpool: Liverpool University Press.

Brooks, D. H. (1965). Root infection by ascospores of *Ophiobolus graminis* as a factor in epidemiology of the take-all disease. *Transactions of the British Mycological Society* **48**, 237–48.

Brown, G. E., and Kennedy, B. W. (1966). Effect of oxygen concentration on *Pythium* seed rot of soybean. *Phytopathology* **56**, 407–11.

Brown, J., Fahim, M. M., and Hutchinson, S. A. (1968). Some effects of atmospheric humidity on the growth of *Serpula lacrimans*. *Transactions of the British Mycological Society* **51**, 507–10.

Brown, J. C. (1958). Soil fungi of some British sand dunes in relation to soil type and succession. *Journal of Ecology* **46**, 641–64.

Bruehl, G. W., and Lai, P. (1968). Influence of soil *p*H and humidity on survival of *Cephalosporium gramineum* in infested wheat straw. *Canadian Journal of Plant Science* **48**, 245–52.

Bruehl, G. W., and Lai, P. (1966). Prior-colonization as a factor in the saprophytic survival of several fungi in wheat straw. *Phytopathology* **56**, 766–8.

Bruehl, G. W., Sprague, R., Fischer, W. R., Nagamitsu, M., Nelson, W. L., and Vogel, O. H. (1966). Snow molds of winter wheat in Washington. *Washington Agricultural Experimental Station, Bulletin* **677**, 1–21.

Bumbieris, M., and Lloyd, A. B. (1967). Influence of soil fertility and moisture on lysis of fungal hyphae. *Australian Journal of Biological Sciences* **20**, 103–12.

Bunt, J. S., and Rovira, A. D. (1955). The effect of temperature and heat treatment on soil metabolism. *Journal of Soil Science* **6**, 129–36.

Burges, A. (1950). The downward movement of fungal spores in sandy soil. *Transactions of the British Mycological Society* **33**, 142–7.

Burges, A. (1958). *Micro-organisms in the Soil.* London: Hutchinson.

Burges, A., and Fenton, E. (1953). The effect of carbon dioxide on the growth of certain soil fungi. *Transactions of the British Mycological Society* **36**, 104–8.

Burges, A., and Nicholas, D. P. (1961). Use of soil sections in studying amount of fungal hyphae in soil. *Soil Science* **92**, 25–9.

Burges, A., and Raw, F. (1967). *Soil Biology.* London: Academic Press.

Burgess, L. W., and Griffin, D. M. (1967). Competitive saprophytic colonization of wheat straw. *Annals of Applied Biology* **60**, 137–42.

Burgess, L. W., and Griffin, D. M. (1968*a*). The recovery of *Gibberella* *zeae* from wheat straws. *Australian Journal of Experimental Agriculture and Animal Husbandry* **8**, 364–70.

Burgess, L. W., and Griffin, D. M. (1968*b*). The influence of diurnal temperature fluctuations on the growth of fungi. *New Phytologist* **67**, 131–7.

Burke, D. W. (1965). The near immobility of *Fusarium solani* f. *phaseoli* in natural soils. *Phytopathology* **55**, 1188–90.

Burke, D. W. (1968). Root growth obstructions and *Fusarium* root rot of beans. *Phytopathology* **58**, 1575–6.

Burke, D. W. (1969). Significance of the plowed soil layer in *Fusarium* root rot of beans. *Phytopathology* **59**, 1020.

Burke, D. W., Hagedorn, D. J., and Mitchell, J. E. (1970). Soil conditions and distribution of pathogens in relation to pea root rot in Wisconsin soils. *Phytopathology* **60**, 403–6.

Buston, H. W., Moss, M. O., and Tyrrell, D. (1966). The influence of carbon dioxide on growth and sporulation of *Chaetomium globosum*. *Transactions of the British Mycological Society* **49**, 387–96.

Butler, F. C. (1953). Saprophytic behaviour of some cereal root-rot fungi. I. Saprophytic colonization of wheat straw. *Annals of Applied Biology* **40**, 284–97.

Caldwell, J. (1963). Effects of high partial pressures of oxygen on fungi. *Nature, London* **197**, 772–4.

Calvet, R. (1967*a*). La diffusion dans les systèmes argile-eau. I. Rappels théoretique concernant les phenomenes de diffusion. *Annales Agronomiques, Paris* **18**, 217–36.

Calvet, R. (1967*b*). La diffusion dans les systems argile-eau. II. Diffusion des cations. *Annales Agronomiques, Paris* **18**, 429–44.

Campbell, A. H. (1933). Zone lines in plant tissues. I. The black lines formed by *Xylaria polymorpha* (Pers.) Grev. in hardwood. *Annals of Applied Biology* **20**, 123–45.

Campbell, A. H. (1934). Zone lines in plant tissues. II. The black lines formed by *Armillaria mellea* (Vahl.) Quel. *Annals of Applied Biology* **21**, 1–22.

Cantino, E. C. (1966). Morphogenesis in aquatic fungi. In *The Fungi*, Vol. 2, G. C. Ainsworth and A. S. Sussman, eds., pp. 283–337. New York: Academic Press.

Casida, L. E. (1969). Observations of microorganisms in soil and other natural habitats. *Applied Microbiology* **18**, 1065–71.

Chapman, H. D. (1965). Chemical factors of the soil as they affect microorganisms. In *Ecology of Soil-borne Plant Pathogens*, K. F.

Baker and W. C. Snyder, eds., pp. 120–39. Berkeley: University of California Press.

Chen, A. W.-C. (1964). Soil fungi with high salt tolerance. *Transactions of the Kansas Academy of Science* 67, 36–40.

Chen, A. W.-C., and Griffin, D. M. (1966a). Soil physical factors and the ecology of fungi. V. Further studies in relatively dry soils. *Transactions of the British Mycological Society* 49, 419–26.

Chen, A. W.-C., and Griffin, D. M. (1966b). Soil physical factors and the ecology of fungi. VI. Interaction between temperature and soil moisture. *Transactions of the British Mycological Society* 49, 551–61.

Chesters, C. G. C., and Thornton, R. H. (1956). A comparison of techniques for isolating soil fungi. *Transactions of the British Mycological Society* 39, 301–13.

Chet, I. (1969). The role of sclerotial rind in the germinability of sclerotia of *Sclerotium rolfsii*. *Canadian Journal of Botany* 47, 593–5.

Chet, I., Henis, Y., and Kislev, N. (1969). Ultrastructure of sclerotia and hyphae of *Sclerotium rolfsii* Sacc. *Journal of General Microbiology* 57, 143–8.

Chet, I., Henis, Y., and Mitchell, R. (1967). Chemical composition of hyphal and sclerotial walls of *Sclerotium rolfsii*. *Canadian Journal of Microbiology* 13, 137–41.

Childs, T. W. (1963). *Poria weirii* root rot. *Phytopathology* 53, 1124–7.

Christensen, C. M., and Kaufmann, H. H. (1965). Deterioration of stored grains by fungi. *Annual Review of Phytopathology* 3, 69–84.

Christensen, M. (1969). Soil microfungi of dry to mesic conifer-hardwood forests in northern Wisconsin. *Ecology* 50, 9–27.

Christensen, M., and Whittingham, W. F. (1965). The soil microfungi of open bogs and conifer swamps in Wisconsin. *Mycologia* 57, 882–96.

Christian, J. H. B., and Waltho, J. A. (1962). The water relations of staphylococci, and micrococci. *Journal of Applied Bacteriology* 25, 369–77.

Christian, J. H. B., and Waltho, J. A. (1964). The composition of *Staphylococcus aureus* in relation to the water activity of the growth medium. *Journal of General Microbiology* 35, 205–13.

Clark, F. E. (1965). The concept of competition in microbial ecology. In *Ecology of Soil-borne Plant Pathogens*, K. F. Baker and W. C. Snyder, eds., pp. 339–45. Berkeley: University of California Press.

Clark, F. E. (1967). Bacteria in soil. In *Soil Biology*, A Burges and F. Raw, eds., pp. 15–49. London: Academic Press.

Clark, F. E. (1968). The growth of bacteria in soil. In *Ecology of Soil Bacteria*, T. R. G. Gray and D. Parkinson, eds., pp. 441–57. Liverpool: Liverpool University Press.

Clark, F. E., Jackson, R. D., and Gardner, H. R. (1962). Measurement of microbial thermogenesis in soil. *Proceedings of the Soil Science Society of America* **26**, 155–60.

Cochrane, V. W. (1958). *Physiology of Fungi*. New York: John Wiley.

Cochrane, V. W. (1966). Respiration and spore germination. In *The Fungus Spore*, Colston Papers no. 18, M. F. Madelin, ed., pp. 201–13. London: Butterworth.

Coley-Smith, J. R. (1960). Studies of the biology of *Sclerotium cepivorum* Berk. IV. Germination of sclerotia. *Annals of Applied Biology* **48**, 8–18.

Coley-Smith, J. R., and King, J. E. (1969). The production by species of *Allium* of alkyl sulphides and their effect on germination of sclerotia of *Sclerotium cepivorum* Berk. *Annals of Applied Biology* **64**: 289–301.

Coley-Smith, J. R., King, J. E., Dickinson, D. J. and Holt, R. W. (1967). Germination of sclerotia of *Sclerotium cepivorum*. *Annals of Applied Biology* **60**, 109–15.

Colman, E. A. (1947). A laboratory procedure for determining the field capacity of soils. *Soil Science* **63**, 277–83.

Cook, R. J., and Bruehl, G. W. (1968). Relative significance of parasitism versus saprophytism in colonization of wheat straw by *Fusarium roseum* 'Culmorum' in the field. *Phytopathology* **58**, 306–8.

Cook, R. J., and Flentje, N. T. (1967). Chlamydospore germination and germling survival of *Fusarium solani* f. *pisi* in soil as affected by soil water and pea seed exudation. *Phytopathology* **57**, 178–82.

Cook, R. J., and Papendick, R. I. (1970a). Soil water potential as a factor in the ecology of *Fusarium roseum* f. sp. *cerealis* 'Culmorum'. *Plant and Soil* **32**, 131–45.

Cook, R. J., and Papendick, R. I. (1970b). Effect of soil water on microbial antagonism and nutrient availability in relation to soil-borne fungal diseases of plants. In *Root Diseases and Soil-borne Pathogens*, T. A. Toussoun, R. V. Bega, and P. E. Nelson, eds., pp. 81–8. Berkeley: University of California Press.

Cook, R. J., and Schroth, M. N. (1965). Carbon and nitrogen compounds and germination of chlamydospores of *Fusarium solani* f. *phaseoli*. *Phytopathology* **55**, 254–6.

Cook, R. J., and Snyder, W. C. (1965). Influence of host exudates on growth and survival of germlings of *Fusarium solani* f. *phaseoli* in soil. *Phytopathology* **55**, 1021–5.

Cooney, D. G., and Emerson, R. (1964). *Thermophilic Fungi*. San Francisco: Freeman.

Covey, R. P. (1970). Effect of oxygen tension on the growth of *Phytophthora cactorum*. *Phytopathology* **60**, 358–9.

Currie, J. A. (1961*a*). Gaseous diffusion in the aeration of aggregated soils. *Soil Science* **92**, 40–5.

Currie, J. A. (1961*b*). Gaseous diffusion in porous media. III. Wet granular materials. *British Journal of Applied Physics* **12**, 275–81.

Curtis, P. J. (1969). Anaerobic growth of fungi. *Transactions of the British Mycological Society* **53**, 299–302.

Darby, R. T., and Goddard, D. R. (1950). Studies of the respiration of the mycelia of the fungus *Myrothecium verrucaria*. *American Journal of Botany* **37**, 379–87.

Day, P. R. (1942). The moisture potential of soils. *Soil Science* **54**, 391–400.

Day, P. R., Bolt, G. H., and Anderson, D. M. (1967). Nature of soil water. In *Irrigation of Agricultural Lands*, Agronomy Monograph 11, R. M. Hagan, H. W. Haise, and T. W. Edminster, eds., pp. 193–208. New York: Academic Press.

Dayal, R., and Barron, G. L. (1970). *Verticillium psalliotae* as a parasite of *Rhopalomyces*. *Mycologia* **62**, 826–30.

Debano, L. F. (1969). Water repellent soils: a worldwide concern in management of soil and vegetation. *Agricultural Science Review* **7**, no. 2, 11–18.

Deverall, B. J. (1965). The physical environment for fungal growth. I. Temperature. In *The Fungi*, Vol. 1, G. C. Ainsworth and A. S. Sussman, eds., pp. 543–50. New York: Academic Press.

Dick, M. W. (1968). Considerations of the role of water on the taxonomy and ecology of the filamentous biflagellate fungi in littoral zones. *Veröffentlichungen des Institut für Meeresforschung in Bremerhaven* **3**, 27–38.

Dickinson, C. H., and Parkinson, D. (1970). Effects of mechanical shaking and water tension on survival and distribution of fungal inoculum in glass microbead media. *Canadian Journal of Microbiology* **16**, 549–52.

Dobbs, C. G., and Hinson, W. H. (1960). Some observations on fungal spores in soil. In *The Ecology of Soil Fungi*, D. Parkinson and J. S. Waid, eds., pp. 33–42. Liverpool: Liverpool University Press.

Dommergues, Y. (1962). Contribution à l'étude de la dynamique micro-bienne des sols en zone semi-aride et en zone tropicale sèche. *Annales Agronomiques, Paris* **13**, 265–324, 391–468.

Domsch, K. H. (1960). Das Pilzspektrum einer Bodenprobe. II. Nach-weis physiologischer Merkmale. *Archiv für Mikrobiologie* **35**, 229–47.

Domsch, K. H., Gams, W., and Weber, E. (1968). Der Einfluss Verschiedener Vorfrüchte auf das Bodenpilzspektrum in Weizen-feldern. *Zeitschrift fur Pflanzenernährung, Düngung, Bodenkunde* **119**, 134–49.

Domsch, K. H., and Schwinn, F. J. (1965). Nachweiss and Isolierung von pflanzenpathogenen Bodenpilzen mit selektiven Vehrfahren. *Zentralblatt für Bakteriologie, Parasitenkunde, Infectionskrankheiten und Hygiene*, Supplementheft **1**, 461–85.

Dubé, A. J., Dodman, B. L., and Flentje, N. T. (1971). The influence of water activity on the growth of *Rhizoctonia solani*. *Australian Journal of Biological Sciences* **24**, 57–65.

Dubey, H. D. (1968). Effect of soil moisture levels on nitrification. *Canadian Journal of Microbiology* **14**, 1348–50.

Dukes, P. D., and Apple, J. L. (1965). Effect of oxygen and carbon dioxide tensions on growth and inoculum potential of *Phytophthora parasitica* var. *nicotianae*. *Phytopathology* **55**, 666–9.

Durbin, R. D. (1959). Factors affecting the vertical distribution of *Rhizoctonia solani*, with special reference to $CO_2$ concentration. *American Journal of Botany* **46**, 22–5.

Durbin, R. D. (1961). Techniques for the observation and isolation of soil microorganisms. *Botanical Review* **27**, 522–60.

Durrell, L. W., and Shields, L. M. (1960). Fungi isolated in culture from soils of the Nevada Test site. *Mycologia* **52**, 636–45.

Dwivedi, R. A., and Garrett, S. D. (1968). Fungal competition in agar plate colonization from soil inocula. *Transactions of the British Mycological Society* **51**, 95–101.

Egeberg, R. O., Elconin, A. F., and Egeberg, M. C. (1964). Effect of salinity and temperature on *Coccidioides immitis* and three antagonistic soil saprophytes. *Journal of Bacteriology* **88**, 473–6.

Eggins, H. O. W., and Malik, K. A. (1969). The occurrence of thermo-philic cellulolytic fungi in a pasture land soil. *Antonie van Leeuwenhock Journal of Microbiology and Serology* **35**, 178–84.

Eicker, A. (1970). Vertical distribution of fungi in Zululand soils. *Transactions of the British Mycological Society* **55**, 45–57.

Ekundayo, J. A. (1966). Further studies on germination of sporangiospores of *Rhizopus arrhizus*. *Journal of General Microbiology* **42**, 283–91.

Ekundayo, J. A., and Carlile, M. J. (1964). The germination of sporangiospores of *Rhizopus arrhizus*; spore swelling and germ-tube emergence. *Journal of General Microbiology* **35**, 261–9.

Elconin, A. F., Egeberg, R. O., and Egeberg, M. C. (1964). Significance of soil salinity on the ecology of *Coccidioides immitis*. *Journal of Bacteriology* **87**, 500–3.

Elliott, H. J. (1970). The role of millipedes in the decomposition of *Pinus radiata* litter in the Australian Capital Territory. *Australian Forest Research* **4**, 3–10.

Emerson, R. (1968). Thermophiles. In *The Fungi*, Vol. 3, G. C. Ainsworth and A. S. Sussman, eds., pp. 105–28. New York: Academic Press.

Emerson, W. W. (1959). The structure of soil crumbs. *Journal of Soil Science* **10**, 235–44.

Emerson, W. W., and Bond, R. D. (1963). The rate of water entry into dry sand and calculation of the advancing contact angle. *Australian Journal of Soil Research* **1**, 9–16.

English, M. P. (1963). Saprophytic growth of keratinophilic fungi on keratin. *Sabouraudia* **2**, 115–30.

English, M. P. (1965). The saprophytic growth of non-keratinophilic fungi on keratinized substrata, and a comparison with keratinophilic fungi. *Transactions of the British Mycological Society* **48**, 219–36.

English, M. P. (1968). The developmental morphology of the perforating organs and eroding mycelium of dermatophytes. *Sabouraudia* **6**, 218–27.

Eren, J., and Pramer, D. (1966). Application of immunofluorescent staining to studies of the ecology of soil micro-organisms. *Soil Science* **101**, 39–45.

Eren, J., and Pramer, D. (1968). Use of a fluorescent brightener as aid to studies of fungistasis and nematophagous fungi in soil. *Phytopathology* **58**, 644–6.

Estermann, E. F., and McLaren, A. D. (1959). Stimulation of bacterial proteolysis by adsorbents. *Journal of Soil Science* **10**, 64–78.

Falk, M., Hartmann, K. A., and Lord, R. C. (1962). Hydration of deoxyribonucleic acid. I. Gravitational study. *Journal of the American Chemical Society* **84**, 3843–6.

Falk, M., Hartmann, K. A., and Lord, R. C. (1963*a*). Hydration of deoxyribonucleic acid. II. Infrared study. *Journal of the American Chemical Society* **85**, 387–91.

Falk, M., Hartmann, K. A., and Lord, R. C. (1963*b*). Hydration of deoxyribonucleic acid. III. A spectroscopic study of the effect of hydration on the structure of deoxyribonucleic acid. *Journal of the American Chemical Society* **85**, 391–4.

Farrell, D. A., and Graecen, E. L. (1966). Resistance to penetration of fine probes in compressible soil. *Australian Journal of Soil Research* **4**, 1–17.

Fawcett, R. G., and Collis-George, N. (1967). A filter-paper method for determining the moisture characteristic of soil. *Australian Journal of Experimental Agriculture and Animal Husbandry* **7**, 162–7.

Fellows, H. (1928). The influence of oxygen and carbon dioxide on the growth of *Ophiobolus graminis*. *Journal of Agricultural Research* **37**, 349–55.

Finstein, M. S., and Alexander, M. (1962). Competition for carbon and nitrogen between *Fusarium* and bacteria. *Soil Science* **94**, 334–9.

Flentje, N. T. (1964). Pre-emergence rotting of peas in South Australia. II. Factors associated with the soil. *Australian Journal of Biological Sciences* **17**, 651–64.

Flentje, N. T., and Saksena, H. K. (1964). Pre-emergence rotting of peas in South Australia. III. Host-pathogen interaction. *Australian Journal of Biological Sciences* **17**, 665–75.

Fletcher, J. and Morton, A. G. (1970). Physiology of germination of *Penicillium griseofulvum* conidia. *Transactions of the British Mycological Society* **54**, 65–81.

Flores-Carreon, A., Reyes, E., and Ruiz-Herrera, J. (1969). Influence of oxygen on maltose metabolism by *Mucor rouxii*. *Journal of General Microbiology* **59**, 13–19.

Follstadt, M. N. (1966). Mycelial growth rate and sporulation of *Alternaria tenuis*, *Botrytis cinerea*, *Cladosporium herbarum* and *Rhizopus stolonifer* in low-oxygen atmospheres. *Phytopathology* **56**, 1098–9.

Ford, E. J., Gold, A. H., and Snyder, W. C. (1970*a*). Soil substances inducing chlamydospore formation by *Fusarium*. *Phytopathology* **60**, 124–8.

Ford, E. J., Gold, A. H., and Snyder, W. C. (1970*b*). Induction of chlamydospore formation in *Fusarium solani* by soil bacteria. *Phytopathology* **60**, 479–84.

Forsyth, E. R., and Eaves, C. A. (1969). Ethylene production from Pardee's $CO_2$ buffers. *Physiologia Plantarum* **22**, 1055–8.

Frankland, J. C. (1966). Succession of fungi on decaying petioles of *Pteridium aquilinum*. *Journal of Ecology* **54**, 41–63.

Frankland, J. C. (1969). Fungal decomposition of bracken petioles. *Journal of Ecology* **57**, 25–36.

Fries, N. (1965). The chemical environment for fungal growth. 3. Vitamins and other organic growth factors. In *The Fungi*, Vol. 1, G. C. Ainsworth and A. S. Sussman, eds., pp. 491–524. New York: Academic Press.

Fulton, J. M., Mortimore, C. G., and Hildebrand, A. A. (1961). Note on the relation of soil bulk density to the incidence of *Phytophthora* root and stalk rot of soybeans. *Canadian Journal of Soil Science* **41**, 247.

Galloway, L. D. (1935). The moisture requirements of mold fungi, with special reference to mildew in textiles. *Journal of the Textile Institute* **26**, 123–9.

Gams, W. (1967). Mikroorganismen in der Wurzelregion von Weizen. *Metteilungen aus der Biologischen Bundesanstalt für Land- und Forstwirtschaft, Berlin-Dahlem* **123**, 1–77.

Gams, W., and Domsch, K. H. (1967). Beiträge zur Anwendung der Bodenwaschtechnik für die Isolierung von Bodenpilzen. *Archiv für Mikrobiologie* **58**, 134–44.

Gams, W., and Domsch, K. H. (1969). The spatial and seasonal distribution of microscopic fungi in arable soils. *Transactions of the British Mycological Society* **52**, 301–8.

Garren, K. H. (1964). Isolation procedures influence the apparent make-up of the terrestrial microflora of peanut pods. *Plant Disease Reporter* **48**, 344–8.

Garrett, M. K., and Robinson, P. M. (1969). A stable inhibitor of spore germination produced by fungi. *Archiv für Mikrobiologie* **67**, 370–7.

Garrett, S. D. (1937). Soil conditions and the take-all disease of wheat. II. The relation between soil reaction and soil aeration. *Annals of Applied Biology* **24**, 747–51.

Garrett, S. D. (1940). Soil conditions and the take-all disease of wheat. V. Further experiments on the survival of *Ophiobolus graminis* in infected wheat stubble buried in the soil. *Annals of Applied Biology* **27**, 199–204.

Garrett, S. D. (1951). Ecological groups of soil fungi: a survey of substrate relationships. *New Phytologist* **50**, 149–66.

Garrett, S. D. (1956a). *Biology of Root-infecting Fungi*. Cambridge: Cambridge University Press.

Garrett, S. D. (1956b). Rhizomorph behaviour in *Armillaria mellea* (Vahl) Quel. II. Logistics of infection. *Annals of Botany* N.S. **20**, 193–210.

Garrett, S. D. (1960). Inoculum potential. In *Plant Pathology*, Vol. 3, J. G. Horsfalls and A. E. Dimond eds., pp. 23–56. New York: Academic Press.

Garrett, S. D. (1963). *Soil Fungi and Soil Fertility*. Oxford: Pergamon Press.

Garrett, S. D. (1966a). Spores as propagules of disease. In *The Fungus Spore*, Colston Papers No. 18, M. F. Madelin, ed., pp. 309–18. London: Butterworths.

Garrett, S. D. (1966b). Cellulose-decomposing ability of some cereal foot-rot fungi in relation to their saprophytic survival. *Transactions of the British Mycological Society* **49**, 57–68.

Garrett, S. D. (1967). Effect of nitrogen level on survival of *Ophiobolus graminis* in pure culture on cellulose. *Transactions of the British Mycological Society* **50**, 519–24.

Garrett, S. D. (1970). *Pathogenic Root-Infecting Fungi*. Cambridge: Cambridge University Press.

Gaumann, E. (1950). *Principles of Plant Infection*. London: Crosby Lockwood.

Gerdemann, J. W. (1968). Vesicular – arbuscular mycorrhiza and plant growth. *Annual Review of Phytopathology* **6**, 397–418.

Gerdemann, J. W., and Nicholson, T. H. (1963). Spores of mycorrhizal *Endogone* species extracted from soil by wet sieving and decanting. *Transactions of the British Mycological Society* **46**, 235–44.

Gerlagh, M. (1968). Introduction of *Ophiobolus graminis* into new polders and its decline. *Netherlands Journal of Plant Pathology* **74**, suppl. 2, 1–97.

Gibbs, J. N. (1967). The role of host vigour in the susceptibility of pines to *Fomes annosus*. *Annals of Botany* N.S. **31**, 803–15.

Gilbert, R. G., and Griebel, G. E. (1969). The influence of volatile substances from alfalfa on *Verticillium dahliae* in soil. *Phytopathology* **59**, 1400–4.

Gilbert, R. G., Menzies, J. D., and Griebel, G. E. (1969). The influence of volatiles from alfalfa upon growth and survival of soil micro-organisms. *Phytopathology* **59**, 992–5.

Gilpatrick, J. D. (1969). Role of ammonia in the control of avocado root rot with alfalfa meal soil amendment. *Phytopathology* **59**, 973–8.

Gochenaur, S. E., and Backus, M. P. (1967). Mycoecology of willow and cottonwood lowland communities in southern Wisconsin. II. Soil microfungi in the sand bar willow stands. *Mycologia* 59, 893–901.

Gochenaur, S. E., and Whittingham, W. F. (1967). Mycoecology of willow and cottonwood lowland communities in southern Wisconsin. I. Soil microfungi in the willow-cottonwood forests. *Mycopathologia et Mycologia Applicata* 33, 125–39.

Gottlieb, D., and Tripathi, R. K. (1968). The physiology of swelling phase of spore germination in *Penicillium atrovenetum*. *Mycologia* 60, 571–90.

Gray, T. R. G., and Baxby, P. (1968). Chitin decomposition in soil. II. The ecology of chitinoclastic microorganisms in forest soil. *Transactions of the British Mycological Society* 51, 293–309.

Gray, T. R. G., Baxby, P., Hill, I. R., and Goodfellow, M. (1968). Direct observation of bacteria in soil. In *The Ecology of Soil Bacteria*, T. R. G. Gray and D. Parkinson, eds., pp. 171–92. Liverpool: Liverpool University Press.

Gray, T. R. G., and Parkinson, D. (1968). *The Ecology of Soil Bacteria*. Liverpool: Liverpool University Press.

Green, R. J. (1969). Survival and inoculum potential of conidia and microsclerotia of *Verticillium alboatrum* in soil. *Phytopathology* 59, 874–6.

Greenland, D. J. (1965a). Interaction between clays and organic compounds in soils. I. Mechanisms of interaction between clays and defined organic compounds. *Soils and Fertilizers* 28, 415–25.

Greenland, D. J. (1965b). Interaction between clays and organic compounds in soils. II. Adsorption of soil organic compounds and its effect on soil properties. *Soils and Fertilizers* 28, 521–32.

Greenwood, D. J. (1961). The effect of oxygen concentration on the decomposition of organic materials in soil. *Plant and Soil* 14, 360–76.

Greenwood, D. J. (1962). Nitrification and nitrate dissimilation in soil. II. Effect of oxygen concentration. *Plant and Soil* 17, 378–91.

Greenwood, D. J. (1963). Nitrogen transformations and the distribution of oxygen in soil. *Chemistry and Industry*, 1963, 799–803.

Greenwood, D. J. (1968a). Measurement of microbial metabolism in soil. In *The Ecology of Soil Bacteria*, T. R. G. Gray and D. Parkinson, eds., pp. 138–57. Liverpool: Liverpool University Press.

Greenwood, D. J. (1968b). Carbon dioxide distribution in soil. *Report of the National Vegetable Research Station, Wellesborne, for 1967*, 34–5.

Greenwood, D. J., and Berry, G. (1962). Aerobic respiration in soil crumbs. *Nature, London* 195, 161–3.

Greenwood, D. J., and Goodman, D. (1965). Oxygen diffusion and aerobic respiration in columns of fine crumbs. *Journal of Science of Food and Agriculture* **16**, 152–60.

Greenwood, D. J., and Goodman, D. (1967). Direct measurements of the distribution of oxygen in soil aggregates and in columns of fine soil crumbs. *Journal of Soil Science* **18**, 182–96.

Griffin, D. H. (1965). The interaction of hydrogen ions, carbon dioxide and potassium ion in controlling the formation of resistant sporagia in *Blastocladiella emersonii*. *Journal of General Microbiology* **40**, 13–28.

Griffin, D. M. (1960). Fungal colonization of sterile hair in contact with soil. *Transactions of the British Mycological Society*, **43**, 583–96.

Griffin, D. M. (1963a). Soil moisture and the ecology of soil fungi. *Biological Reviews* **38**, 141–66.

Griffin, D. M. (1963b). Soil physical factors and the ecology of fungi. I. Behaviour of *Curvularia ramosa* at small soil water suctions. *Transactions of the British Mycological Society* **46**, 273–280.

Griffin, D. M. (1963c). Soil physical factors and the ecology of fungi. II. Behaviour of *Pythium ultimum* at small soil water suctions. *Transactions of the British Mycological Society* **46**, 368–72.

Griffin, D. M. (1963d). Soil physical factors and the ecology of fungi. III. Activity of fungi in relatively dry soil. *Transactions of the British Mycological Society* **46**, 373–7.

Griffin, D. M. (1966a). Soil physical factors and the ecology of fungi. IV. Influence of the soil atmosphere. *Transactions of British Mycological Society* **49**, 115–20.

Griffin, D. M. (1966b). Soil water terminology in mycology and plant pathology. *Transactions of the British Mycological Society* **49**, 367–8.

Griffin, D. M. (1966c). Fungi attacking seeds in dry seed-beds. *Proceedings of the Linnean Society of New South Wales* **91**, 84–9.

Griffin, D. M. (1968a). A theoretical study relating the concentration and diffusion of oxygen to the biology of organisms in soil. *New Phytologist* **67**, 561–77.

Griffin, D. M. (1968b). Observations on fungi growing in a translucent particulate matrix. *Transactions of the British Mycological Society* **51**, 319–22.

Griffin, D. M. (1969). Soil water in the ecology of fungi. *Annual Review of Phytopathology* **7**, 289–310.

Griffin, D. M. (1970). Effect of soil moisture and aeration on fungal activity – an introduction. In *Root Diseases and Soil-borne Pathogens*,

M

T. A. Toussoun, R. V. Bega, and P. E. Nelson, eds. pp. 77–80. Berkeley: University of California Press.

Griffin, D. M., and Nair, N. G. (1968). Growth of *Sclerotium rolfsii* at different concentrations of oxygen and carbon dioxide. *Journal of Experimental Botany* **19**, 812–16.

Griffin, D. M., Nair, N. G., Baxter, R. I., and Smiles, D. E. (1967). Control of gaseous environment of organisms using a diffusion column technique. *Journal of Experimental Botany* **18**, 518–25.

Griffin, D. M., and Quail, G. (1968). Movement of bacteria in moist, particulate systems. *Australian Journal of Biological Sciences* **21**, 579–82.

Griffin, G. J. (1970). Carbon and nitrogen requirements for macroconidial germination of *Fusarium solani*: dependence on conidial density. *Canadian Journal of Microbiology* **16**, 733–40.

Griffin, G. J., and Pass, T. (1969). Behaviour of *Fusarium roseum* 'Sambucinum' under carbon starvation conditions in relation to survival in soil. *Canadian Journal of Microbiology* **15**, 117–26.

Griffiths, E. (1965). Micro-organisms and soil structure. *Biological Reviews* **40**, 129–42.

Griffiths, E., and Siddiqi, M. A. (1961). Some factors affecting occurrence of *Fusarium culmorum* in the soil. *Transactions of the British Mycological Society* **44**, 343–53.

Groom, P., and Panniset, T. (1933). Studies on *Penicillium chrysogenum* Thom, in relation to temperature and relative humidity of the air. *Annals of Applied Biology* **20**, 633–60.

Gunner, H. B., and Alexander, M. (1964). Anaerobic growth of *Fusarium oxysporum*. *Journal of Bacteriology* **87**, 1309–16.

Gupta, U. C. (1967). Carbohydrates. In *Soil Biochemistry*, A. D. McLaren and G. F. Peterson, eds., pp. 91–118. New York: Marcel Dekker.

Hack, A. R. B. (1956). An application of a method of gas microanalysis to the study of soil air. *Soil Science* **82**, 217–31.

Harley, J. L. (1969). *The Biology of Mycorrhiza*, 2nd ed. London: Leonard Hill.

Harper, J. E., and Webster, J. (1964). An experimental analysis of the coprophilous fungus succession. *Transactions of the British Mycological Society* **47**, 511–30.

Harris, R. F., Gardner, W. R., Adebayo, A. A., and Sommers, L. E. (1970). Agar dish isopiestic equilibration method for controlling the water potential of solid substrates. *Applied Microbiology* **19**, 536–7.

Hawthorne, B. T., and Tsao, P. H. (1969). Inadequacy of the nutrient hypothesis to explain soil fungistasis in relation to chlamydospores of *Thielaviopsis basicola*. *Phytopathology* **59**, 1030.

Hayes, W. A., Randle, P. E., and Last, F. T. (1969). The nature of the microbial stimulus affecting sporophore formation in *Agaricus bisporus* (Lange) Sing. *Annals of Applied Biology* **64**, 177–87.

Hayman, D. S. (1969). The influence of temperature on the exudation of nutrients from cotton seeds and on pre-emergence damping-off by *Rhizoctonia solani*. *Canadian Journal of Botany* **47**, 1663–9.

Heale, J. B., and Isaac, I. (1963). Wilt of lucerne caused by species of *Verticillium*. IV. Pathogenicity of *V. albo-atrum* and *V. dahliae* to lucerne and other crops; spread and survival of *V. albo-atrum* in soil and in weeds; effect upon lucerne. *Annals of Applied Biology* **52**, 439–51.

Heintzeler, I. (1939). Das Wachstum der Schimmelpilze in Abhangigkeit von der Hydraturverhältnissen unter verscheidenen Aussenbedingungen. *Archiv für Mikrobiologie* **10**, 92–132.

Hendley, N. I. (1964). Some observations on *Cladosporium resinae* as a fuel contaminant and its possible role in the corrosion of aluminium alloy fuel tanks. *Transactions of the British Mycological Society* **47**, 467–75.

Henis, Y., and Ben-Yephet, Y. (1970). Effect of propagule size of *Rhizoctonia solani* on saprophytic growth, infectivity, and virulence on bean seedlings. *Phytopathology* **60**, 1351–6.

Hepple, S. (1960). The movement of fungal spores in soil. *Transactions of the British Mycological Society* **43**, 73–9.

Hickman, C. J., and English, M. P. (1951). Factors influencing the development of red core in strawberries. *Transactions of the British Mycological Society* **34**, 223–36.

Hickman, C. J., and Ho, H. H. (1966). Behaviour of zoospores in plant-pathogenic phycomycetes. *Annual Review of Phytopathology* **4**, 195–220.

Hill, I. R., and Gray, T. R. G. (1967). Application of the fluorescent-antibody technique to an ecological study of bacteria in soil. *Journal of Bacteriology* **93**, 1888–96.

Ho, H. H., and Hickman, C. J. (1967). Asexual reproduction and behaviour of zoospores of *Phytophthora megasperma* var. *sojae*. *Canadian Journal of Botany* **45**, 1963–81.

Holmes, J. W., Taylor, S. A., and Richards, S. J. (1967). Measurement of soil water. In *Irrigation of Agricultural Lands*, Agronomy Monograph 11, R. M. Hagan, H. W. Haise, and T. W. Edminster, eds., pp. 275–303. New York: Academic Press.

Hornby, D. (1969). Methods of investigating populations of the take-all fungus (*Ophiobolus graminis*) in soil. *Annals of Applied Biology* **64**, 503–13.

Hsu, S. C., and Lockwood, J. L. (1969). Mechanisms of inhibition of fungi in agar by *Streptomycetes*. *Journal of General Microbiology* **57**, 149–58.

Huber, D. M., Anderson, A. L., and Finley, A. M. (1966). Mechanisms of biological control in a bean root rot soil. *Phytopathology* **56**, 953–6.

Hudson, H. J. (1968). The ecology of fungi on plant remains above the soil. *New Phytologist* **67**, 837–74.

Hunter, R. E., and Guinn, G. (1968). Effect of root temperature on hypocotyls of cotton seedlings as a source of nutrition for *Rhizoctonia solani*. *Phytopathology* **58**, 981–4.

Hurst, H. M. and Burges, N. A. (1967). Lignin and humic acids. In *Soil Biochemistry*, A. D. McLaren and G. F. Peterson, eds., pp. 260–86. New York: Marcel Dekker.

Hutchinson, S. A., and Kamel, M. (1956). The effect of earthworms on the dispersal of soil fungi. *Journal of Soil Science* **7**, 213–18.

Ikediugwu, F. E. O., Dennis, C., and Webster, J. (1970). Hyphal interference by *Peniophora gigantea* against *Heterobasidion annosum*. *Transactions of the British Mycological Society* **54**, 307–9.

Ikediugwu, F. E. O., and Webster, J. (1970*a*). Antagonism between *Coprinus heptemerus* and other coprophilous fungi. *Transactions of the British Mycological Society* **54**, 181–204.

Ikediugwu, F. E. O., and Webster, J. (1970*b*). Hyphal interference in a range of coprophilous fungi. *Transactions of the British Mycological Society* **54**, 205–10.

Ingold, C. T. (1966). Aspects of spore liberation: violent discharge. In *The Fungus Spore*, Colston Papers No. 18, M. F. Madelin, ed., pp. 113–33. London: Butterworth.

Ingram, M. (1957). Micro-organisms resisting high concentrations of sugars or salts. In *Microbial Ecology*, 7th Symposium of the Society for General Microbiology, R. E. O. Williams and C. C. Spicer, eds., pp. 90–133. Cambridge: Cambridge University Press.

Isaac, I. (1953). A further comparative study of pathogenic isolates of *Verticillium*: *V. nubilum* Pethybr. and *V. tricorpus* sp. nov. *Transactions of the British Mycological Society* **36**, 180–95.

Jackson, R. M. (1965). Antibiosis and fungistasis in soil microorganisms.

In *Ecology of Soil-borne Plant Pathogens*, K. F. Baker and W. C. Snyder, eds., pp. 363–9. Berkeley: University of California Press.

Jensen, H. L. (1934). Contributions to the microbiology of Australian soils. I. Numbers of microorganisms in soil, and their relation to certain external factors. *Proceedings of the Linnean Society of New South Wales* 59, 101–17.

Jensen, K. F. (1967). Oxygen and carbon dioxide affect the growth of wood-decaying fungi. *Forest Science* 13, 384–9.

Jensen, K. F. (1969). Effect of constant and fluctuating temperature on growth of four wood-decaying fungi. *Phytopathology* 59, 645–7.

Johnson, L. F., Curl, E. A., Bond, J. H., and Fribourg, H. A. (1959). *Methods for Studying Soil Microflora–Plant Disease Relationships.* Minneapolis: Burgess Publishing Company.

Johnson, M. J. (1967). Aerobic microbial growth at low oxygen concentrations. *Journal of Bacteriology* 94, 101–8.

Johnson, T. W. (1968). Saprobic marine fungi. In *The Fungi*, Vol. 3, G. C. Ainsworth and A. S. Sussman, eds., pp. 95–104. New York: Academic Press.

Johnson, T. W., and Sparrow, F. K. (1961). *Fungi in Oceans and Estuaries.* Weinheim: J. Cramer.

Jones, D. (1970). Ultrastructure and composition of the cell walls of *Sclerotinia sclerotiorum. Transactions of the British Mycological Society* 54, 351–60.

Jones, D., and Farmer, V. C. (1967). The ecology and physiology of soil fungi involved in the degradation of lignin and related aromatic compounds. *Journal of Soil Science* 18, 74–84.

Jones, D., and Webley, D. M. (1968). A new enrichment technique for studying lysis of fungal cell walls in soil. *Plant and Soil* 28, 147–57.

Jones, P. C. T., and Mollison, J. E. (1948). A technique for the quantitative estimation of soil microorganisms. *Journal of General Microbiology* 2, 54–69.

Jooste, W. J. (1966). The effect of different crop sequences on the rhizosphere fungi of wheat. *South African Journal of Agricultural Science* 9, 127–36.

Jørgensen, J. (1964). Some observations on the effect of temperature on the sporulation of *Cercosparella herpotrichoides* Fron. *Acta Agriculturae Scandinavica* 14, 126–8.

Justice, K. J., and Smith, R. L. (1962). Nitrification of ammonium sulphate in calcareous soil as influenced by combinations of moisture, temperature, and levels of added nitrogen. *Proceedings of the Soil Science Society of America* **26**, 246–50.

Kerr, A. (1964). The influence of soil moisture on infection of peas by *Pythium ultimum*. *Australian Journal of Biological Sciences* **17**, 676–85.

Kevan, D. K. McE. (1962). *Soil Animals*. London: Witherby.

Keynan, A., Henis, Y., and Keller, P. (1961). Factors influencing the composition of the cellulose-decomposing microflora on soil crumb plates. *Nature, London* **191**, 307.

Keyworth, W. G., and Milne, L. J. R. (1969). Induced tolerance of *Sclerotium cepivorum* to antibiotics in the presence of onion exudates. *Annals of Applied Biology* **63**, 415–24.

King, J. E., and Coley-Smith, J. R. (1969*a*). Production of volatile alkyl sulphides by microbial degradation of synthetic alliin and alliin-like compounds in relation to germination of sclerotia of *Sclerotium cepivorum* Berk. *Annals of Applied Biology* **64**, 303–14.

King, J. E., and Coley-Smith, J. R. (1969*b*). Suppression of sclerotial germination in *Sclerotium cepivorum* Berk. by water expressed from four soils. *Soil Biology and Biochemistry* **1**, 83–7.

Klotz, L. J., Stolzy, L. H., and DeWolfe, T. A. (1963). Oxygen requirement of three root-rotting fungi in a liquid medium. *Phytopathology* **53**, 302–5.

Klotz, L. J., Stolzy, L. H., DeWolfe, T. A., and Szuszkiewicz, T. E. (1965). Rate of oxygen supply and distribution of root-rotting fungi in soils. *Soil Science* **99**, 200–4.

Ko, W.-H., and Lockwood, J. L. (1967). Soil fungistasis: Relation to fungal spore nutrition. *Phytopathology* **57**, 894–901.

Ko, W.-H., and Lockwood, J. L. (1970). Mechanism of lysis of fungal mycelia in soil. *Phytopathology* **60**, 148–54.

Kouyeas, V. (1964). An approach to the study of moisture relations of soil fungi. *Plant and Soil* **20**, 351–63.

Kouyeas, V., and Balis, C. (1968). Influence of moisture on the restoration of mycostasis in air dried soils. *Annales de L'Institute Phytopathologique Benaki* N.S. **8**, 123–44.

Kubiena, W. L. (1938). *Micropedology*. Iowa: Collegiate Press.

Kuehn, H. H., Orr, G. F., and Ghosh, G. R. (1964). Pathological implications of Gymnoascaceae. *Mycopathologia et Mycologia Applicata* **24**, 35–46.

Kuo, M.-J., and Alexander, M. (1967). Inhibition of the lysis of fungi by melanins. *Journal of Bacteriology* **94**, 624–9.

Lai, P., and Bruehl, G. W. (1966). Survival of *Cephalosporium gramineum* in naturally infested wheat straws in soil in the field and in the laboratory. *Phytopathology* 56, 213–18.

Lai, P., and Bruehl, G. W. (1968). Antagonism among *Cephalosporium gramineum*, *Trichoderma* spp., and *Fusarium culmorum*. *Phytopathology* 58, 562–6.

Lang, A. R. G. (1967). Osmotic coefficients and water potentials of sodium chloride solutions from 0° to 40°C. *Australian Journal of Chemistry* 20, 2017–23.

Lanigan, G. W. (1963). Silage bacteriology. I. Water activity and temperature relationships of silage strains of *Lactobacillus plantarum*, *Lactobacillus brevis* and *Pediococcus cerevisiae*. *Australian Journal of Biological Sciences* 16, 606–15.

Lapwood, D. H. (1966). The effects of soil moisture at the time potato tubers are forming on the incidence of common scab (*Streptomyces scabies*). *Annals of Applied Biology* 58, 447–56.

Larsen, H. (1967). Biochemical aspects of extreme halophilism. *Advances in Microbial Physiology* 1, 97–132.

Latham, A. J., and Watson, R. D. (1967). Effect of crop residues on soil fungi and onion growth in naturally infested soil. *Phytopathology* 57, 505–9.

Leeper, G. W. (1964). *Introduction to Soil Science*, 4th edition. Melbourne: Melbourne University Press.

Lemon, E. R., and Erickson, A. E. (1952). The measurement of oxygen diffusion in the soil with a platinum electrode. *Proceedings of the Soil Science Society of America* 16, 160–3.

Lemon, E. R., and Erickson, A. E. (1955). Principle of the platinum electrode as a method of characterising soil aeration. *Soil Science* 79, 383–92.

Leonard, O. A. (1945). Cotton root development in relation to natural aeration of some Mississippi blackbelt and delta soils. *Journal of American Society of Agronomy* 37, 55–71.

Letey, J., and Stolzy, L. H. (1964). Measurement of oxygen diffusion rates with the platinum micro-electrode. I. Theory and equipment. *Hilgardia* 35, 545–54.

Levi, M. P., and Cowling, E. B. (1969). Role of nitrogen in wood deterioration. VII. Physiological adaptation of wood-destroying and other fungi to substrates deficient in nitrogen. *Phytopathology* 59, 460–8.

Levi, M. P., Merrill, W., and Cowling, E. B. (1968). Role of nitrogen in wood deterioration. VI. Mycelial fractions and model nitrogen compounds as substrates for growth of *Polyporus versicolor* and other wood-destroying and wood-inhabiting fungi. *Phytopathology* **58**, 626–34.

Lewis, B. G. (1970). Effects of water potential on the infection of potato tubers by *Streptomyces scabies* in soil. *Annals of Applied Biology* **66**, 83–8.

Lewis, J. A., and Starkey, R. L. (1969). Decomposition of plant tannins by some microorganisms. *Soil Science* **107**, 235–41.

Linderman, R. G. (1970). Plant residue decomposition products and their effects on host roots and fungi pathogenic to roots. *Phytopathology* **60**, 19–22.

Linderman, R. G., and Gilbert, R. G. (1969). Stimulation of *Sclerotium rolfsii* in soil by volatile components of alfalfa hay. *Phytopathology* **59**, 1366–72.

Lindsay, D. L. (1965). Ecology of plant pathogens in soil. III. Competition between soil fungi. *Phytopathology* **55**, 104–10.

Lloyd, A. D. (1969). Behaviour of Streptomycetes in soil. *Journal of General Microbiology* **59**, 165–70.

Lloyd, A. D., and Lockwood, J. L. (1966). Lysis of fungal hyphae in soil and its possible relation to autolysis. *Phytopathology* **56**, 595–602.

Lockhart, C. L. (1967a). Effect of temperature and various $CO_2$ and $O_2$ concentrations on growth of *Typhula* sp., a parasitic fungus of strawberry plants. *Canadian Journal of Plant Science* **47**, 450–2.

Lockhart, C. L. (1967b). Influence of controlled atmospheres on the growth of *Gloeosporium album* in vitro. *Canadian Journal of Plant Science* **47**, 649–51.

Lockhart, C. L. (1968). Influence of various carbon dioxide and oxygen concentrations on the growth of *Fusarium oxysporum* in vitro. *Canadian Journal of Plant Science* **48**, 451–3.

Lockwood, J. L. (1960). Lysis of mycelium of plant-pathogenic fungi by natural soil. *Phytopathology* **50**, 787–9.

Lockwood, J. L. (1964). Soil fungistasis. *Annual Review of Phytopathology* **2**, 341–62.

Louvet, J., and Bulit, J. (1964). Recherches sur l'écologie des champignons parasites dans le sol. I. Action du gaz carbonique sur la croissance et l'activité parasitaire de *Sclerotinia minor* et de *Fusarium oxysporum* f. *melonis*. *Annales de Epiphyties* **15**, 21–44.

Macauley, B. J., and Griffin, D. M. (1969a). Effects of carbon dioxide and oxygen on the activity of some soil fungi. *Transactions of the British Mycological Society* 53, 53–62.

Macauley, B. J., and Griffin, D. M. (1969*b*). Effect of carbon dioxide and the bicarbonate ion on the growth of some soil fungi. *Transactions of the British Mycological Society* 53, 223–8.

Macfayden, A. (1970). Simple methods for measuring and maintaining the proportion of carbon dioxide in air, for use in ecological studies of soil respiration. *Soil Biology and Biochemistry* 2, 9–18.

Mandels, M., and Reese, E. T. (1965). Inhibition of cellulases. *Annual Review of Phytopathology* 3, 85–102.

Marchant, R., and White, M. F. (1966). Spore swelling and germination in *Fusarium culmorum*. *Journal of General Microbiology* 42, 237–44.

Marchant, R., and White, M. F. (1967). The carbon metabolism and swelling of *Fusarium culmorum* conidia. *Journal of General Microbiology* 48, 65–77.

Martín, J. F., and Nicolás, G. (1970). Physiology of spore germination in *Penicillium notatum* and *Trichoderma lignorum*. *Transactions of the British Mycological Society* 55, 141–8.

Marx, D. H. (1969*a*). The influence of ectotrophic mycorrhizal fungi on the resistance of pine roots to pathogenic infections. I. Antagonism of mycorrhizal fungi to root pathogenic fungi and soil bacteria. *Phytopathology* 59, 153–63.

Marx, D. H. (1969*b*). The influence of ectotrophic mycorrhizal fungi on the resistance of pine roots to pathogenic infections. II. Production, identification and biological activity of antibiotics produced by *Leucopaxillus cerealis* var. *piceina*. *Phytopathology* 59, 411–17.

Marx, D. H. (1970). The influence of ectotrophic mycorrhizal fungi on the resistance of pine roots to pathogenic infections. V. Resistance of mycorrhizae to infection by vegetative mycelium of *Phytophthora cinnamomi*. *Phytopathology* 60, 1472–3.

Marx, D. H., and Bryan, W. C. (1969). Effect of soil bacteria on the mode of infection of pine roots by *Phytophthora cinnamomi*. *Phytopathology* 59, 614–19.

Marx, D. H., and Davey, C. B. (1969*a*). The influence of ectotrophic mycorrhizal fungi on the resistance of pine roots to pathogenic infections. III. Resistance of aseptically formed mycorrhizae to infection by *Phytophthora cinnamomi*. *Phytopathology* 59, 549–58.

Marx, D. H., and Davey, C. B. (1969*b*). The influence of ectotrophic mycorrhizal fungi on the resistance of pine roots to pathogenic infections. IV. Resistance of naturally occurring mycorrhizae to infections by *Phytophthora cinnamomi*. *Phytopathology* **59**, 559–65.

Matthews, S., and Whitbread, R. (1968). Factors influencing pre-emergence mortality in peas. I. An association between seed exudates and the incidence of pre-emergence mortality in wrinkle-seeded peas. *Plant Pathology* **17**, 11–17.

McIntyre, D. S. (1966*a*). Characterizing soil aeration with a platinum micro-electrode. I. Response in relation to field moisture conditions and electrode diameter. *Australian Journal of Soil Research* **4**, 95–102.

McIntyre, D. S. (1966*b*). Characterizing soil aeration with a platinum micro-electrode. II. Response under controlled soil conditions. *Australian Journal of Soil Research* **4**, 103–14.

McIntyre, D. S. (1967). Physical factors affecting operation of the oxygen cathode in unsaturated porous media. *Soil Science* **103**, 118–25.

McLaren, A. D., and Skujins, J. (1967). The physical environment of microorganisms in soil. In *The Ecology of Soil Bacteria*, T. R. G. Gray and D. Parkinson, eds., pp. 3–24. Liverpool: Liverpool University Press.

Menzies, J. D. (1963*a*). The direct assay of plant pathogen populations in soil. *Annual Review of Phytopathology* **1**, 127–42.

Menzies, J. D. (1963*b*). Survival of microbial plant pathogens in soil. *Botanical Review* **29**, 79–122.

Menzies, J. D., and Gilbert, R. G. (1967). Responses of the soil microflora to volatile components in plant residues. *Proceedings of the Soil Science Society of America* **31**, 495–6.

Merrill, W., and Cowling, E. B. (1966). Role of nitrogen in wood deterioration: Amount and distribution of nitrogen in fungi. *Phytopathology* **56**, 1083–90.

Milburn, J. A. (1970). Cavitation and osmotic potential of *Sordaria* ascospores. *New Phytologist* **69**, 133–41.

Miller, R. D., and Johnson, D. D. (1964). The effect of soil moisture tension on carbon dioxide evolution, nitrification and nitrogen mineralization. *Proceedings of the Soil Science Society of America* **28**, 644–7.

Mircetich, S. M., and Keil, H. L. (1970). *Phytophthora cinnamomi* root rot and stem canker of peach trees. *Phytopathology* **60**, 1376–82.

Moser, U. S., and Olsen, R. V. (1953). Sulfur oxidation in four soils as influenced by soil moisture tension and sulfur bacteria. *Soil Science* **76**, 251–7.

Moubasher, A. H., and El-Dohlob, S. M. (1970). Seasonal fluctuations of Egyptian soil fungi. *Transactions of the British Mycological Society* **54**, 45–51.

Mozumder, B. K. G., and Caroselli, N. E. (1966). The influence of substrate moisture on the growth of *Verticillium albo-atrum* R. and B. *Advancing Frontiers of Plant Science* **16**, 77–83.

Mozumder, B. K. G., and Caroselli, N. E. (1970). Water relations of respiration of *Verticillium albo-atrum* conidia. *Phytopathology* **60**, 915–16.

Nair, N. G., White, N. H., Griffin, D. M., and Blair, S. (1969). Fine structure and electron cytochemical studies of *Sclerotium rolfsii* Sacc. *Australian Journal of Biological Sciences* **22**, 639–52.

Nash, S. M., and Alexander, J. V. (1965). Comparative survival of *Fusarium solani* f. *cucurbitae* and *F. solani* f. *phaseoli* in soil. *Phytopathology* **55**, 963–6.

Newcombe, M. (1960). Some effects of water and anaerobic conditions on *Fusarium oxysporum* f. *cubense* in soil. *Transactions of the British Mycological Society* **43**, 51–9.

Newman, E. I. (1969a). Resistance to water flow in soil and plant. I. Soil resistance in relation to amounts of root: theoretical estimates. *Journal of Applied Ecology* **6**, 1–12.

Newman, E. I. (1969b). Resistance to water flow in soil and plant. II. A review of experimental evidence on the rhizosphere resistance. *Journal of Applied Ecology* **6**, 261–72.

Nicholas, D. J. D. (1965). Utilization of inorganic nitrogen compounds and amino acids by fungi. In *The Fungi*, Vol. 2, G. C. Ainsworth and A. S. Sussman, eds., pp. 349–76. New York: Academic Press.

Nicholas, D. P., Parkinson, D., and Burges, N. A. (1965). Studies of fungi in a podsol. II. Application of the soil-sectioning technique to the study of amounts of fungal mycelium in the soil. *Journal of Soil Science* **16**, 258–69.

Nicot, J. (1960). Some characteristics of the microflora in desert sands. In *The Ecology of Fungi*, D. Parkinson and J. S. Waid, eds., pp. 94–7. Liverpool: Liverpool University Press.

Norstadt, F. A., and McCalla, T. M. (1969). Microbial populations in stubble-mulched soil. *Soil Science* **107**, 188–93.

Nyvall, R. F., and Kommedahl, T. (1970). Saprophytism and survival of *Fusarium moniliforme* in corn stalks. *Phytopathology* **60**, 1233–5.

Old, K. M. (1967). Effects of natural soil on survival of *Cochliobolus sativus*. *Transactions of the British Mycological Society* **50**, 615–24.

Old, K. M. (1969). Perforation of conidia of *Cochliobolus sativus* in natural soils. *Transactions of the British Mycological Society* **53**, 207–16.

Old, K. M., and Robertson, W. M. (1969). Examination of conidia of *Cochliobolus sativus* recovered from natural soil using transmission and scanning electron microscopy. *Transactions of the British Mycological Society* **53**, 217–21.

Old, K. M., and Robertson, W. M. (1970a). Growth of bacteria within lysing fungal conidia in soil. *Transactions of the British Mycological Society* **54**, 337–41.

Old, K. M., and Robertson, W. M. (1970b). Effects of lytic enzymes and natural soil on the fine structure of conidia of *Cochliobolus sativus*. *Transactions of the British Mycological Society* **54**, 343–50.

Owens, L. D., Gilbert, R. G., Griebel, G. E., and Menzies, J. D. (1969). Identification of plant volatiles that stimulate microbial respiration and growth in soil. *Phytopathology* **59**, 1468–72.

Papavizas, G. C. (1968). Survival of root-infecting fungi in soil. VIII. Distribution of *Rhizoctonia solani* in various physical fractions of naturally and artificially infested soils. *Phytopathology* **58**, 746–51.

Papavizas, G. C., and Davey, C. B. (1962). Activity of *Rhizoctonia* in soil as affected by carbon dioxide. *Phytopathology* **52**, 759–65.

Parbery, D. G. (1969). The natural occurrence of *Cladosporium resinae*. *Transactions of the British Mycological Society*, **53**, 15–23.

Park, D. (1959). Some aspects of the biology of *Fusarium oxysporum* Schl. in soil. *Annals of Botany* N.S. **23**, 35–50.

Park, D. (1960). Antagonism – the background to soil fungi. In *The Ecology of Soil Fungi*, D. Parkinson and J. S. Waid, eds., pp. 148–59. Liverpool: Liverpool University Press.

Park, D. (1961a). Isolation of *Fusarium oxysporum* from soils. *Transactions of the British Mycological Society* **44**, 119–22.

Park, D. (1961b). Morphogenesis, fungistasis and cultural staling in *Fusarium oxysporum*, Snyder and Hansen. *Transactions of the British Mycological Society* **44**, 377–90.

Park, D. (1963). The presence of *Fusarium oxysporum* in soils. *Transactions of the British Mycological Society* **46**, 444–8.

Park, D. (1967). The importance of antibiotics and inhibiting substances. In *Soil Biology*, A. Burges and F. Raw, eds., pp. 435–47. London: Academic Press.

Park, D., and Robinson, P. M. (1966). Internal pressure of hyphal tips of fungi, and its significance in morphogenesis. *Annals of Botany* N.S. 30, 425–39.

Parkinson, D. (1967). Soil micro-organisms and plant roots. In *Soil Biology*, A. Burges and F. Raw, eds., pp. 449–78. London: Academic Press.

Parkinson, D., and Balasooriya, I. (1969). Studies on fungi in a pinewood soil. IV. Seasonal and spatial variations in the fungal populations. *Revue d'Ecologie et Biologie du Sol* 6, 147–53.

Parkinson, D., and Williams, S. T. (1961). A method for isolating fungi from soil micro-habitats. *Plant and Soil* 13, 347–55.

Parmeter, J. R., and Hood, J. R. (1962). Use of ultraviolet light in isolation of certain fungi from soil. *Phytopathology* 52, 376–7.

Parr, J. F., and Norman, A. G. (1964). Growth and activity of soil microorganisms in glass micro-beads: I. Carbon dioxide evolution. *Soil Science* 97, 361–6.

Parr, J. F., Parkinson, D., and Norman, A. G. (1963). A glass microbead system for the investigation of soil micro-organisms. *Nature, London* 200, 1227–8.

Parr, J. F., Parkinson, D., and Norman, A. G. (1967). Growth and activity of soil microorganisms in glass micro-beads. II. Oxygen uptake and direct observations. *Soil Science* 103, 303–10.

Penman, H. L. (1940). Gas and vapour movements in the soil. I. The diffusion of vapours through porous solids. *Journal of Agricultural Science* 30, 437–61.

Pentland, G. D. (1967). The effect of soil moisture on the growth and spread of *Coniophora puteana* under laboratory conditions. *Canadian Journal of Botany* 45, 1899–1906.

Persson-Hüppel, A. (1963). The influence of temperature on the antagonistic effect of *Trichoderma viride* Fr. on *Fomes annosus*. *Studia Forestalia Suecica* 4, 1–13.

Peyronel, B. (1956). Considerazioni sulle micocenosi e sui metodi per studiarle. *Allionia* 3, 85–109.

Phillips, D. H. (1966). Oxygen transfer into mycelial pellets. *Biotechnology and Bioengineering* 8, 456–60.

Pirozynski, K. A. (1968). Geographical distribution of fungi. In *The Fungi*, Vol. 3, G. C. Ainsworth and A. S. Sussman, eds., pp. 487–504. New York: Academic Press.

Pirt, S. J. (1967). A kinetic study of the mode of growth of surface colonies of bacteria and fungi. *Journal of General Microbiology* 47, 181–97.

Pitt, J. I., and Christian, J. H. B. (1968). Water relations of xerophilic fungi isolated from prunes. *Applied Microbiology* **16**, 1853–8.

Poel, L. W. (1960). The estimation of oxygen diffusion rates in soils. *Journal of Ecology* **48**, 165–73.

Ponnamperuma, F. N. (1967). A theoretical study of aqueous carbonate equilibria. *Soil Science* **103**, 90–100.

Ponnamperuma, F., Martinez, E., and Loy, T. (1966). Influence of redox potential and partial pressure of carbon dioxide on pH values and the suspension effect of flooded soils. *Soil Science* **101**, 421–31.

Poole, T. B. (1959). Studies on the food of Collembola in a Douglas Fir plantation. *Proceedings of the Zoological Society of London* **132**, 71–82.

Potgieter, H. J., and Alexander, M. (1966). Susceptibility and resistance of several fungi to microbial lysis. *Journal of Bacteriology* **91**, 1526–33.

Pugh, G. J. F. (1962). Studies on fungi in coastal soils. II. Fungal ecology in a developing salt marsh. *Transactions of the British Mycological Society* **45**, 560–6.

Rashevsky, N. (1960). *Mathematical Biophysics*. New York: Dover.

Richards, B. G. (1969). Psychrometric techniques for measuring soil water potential. *Commonwealth Scientific and Industrial Research Organization, Australia, Division of Soil Mechanics Technical Report* No. 9, 1–32.

Rishbeth, J. (1951). Observations on the biology of *Fomes annosus* with particular reference to East Anglian pine plantations. II. Spore production, stump infection, and saprophytic activity in stumps. *Annals of Botany* N.S. **15**, 1–21.

Rishbeth, J. (1955). Fusarium wilt of bananas in Jamaica. I. Some observations on the epidemiology of the disease. *Annals of Botany* N.S. **19**, 293–328.

Rishbeth, J. (1957). Fusarium wilt of bananas in Jamaica. II. Some aspects of host-parasite relationships. *Annals of Botany* N.S. **21**, 215–45.

Rishbeth, J. (1968). The growth rate of *Armillaria mellea*. *Transactions of the British Mycological Society* **51**, 575–86.

Ritchie, D. (1957). Salinity optima for marine fungi affected by temperature. *American Journal of Botany* **44**, 870–4.

Ritchie, D. (1959). The effect of salinity and temperature on marine and other fungi from various climates. *Bulletin of the Torrey Botanical Club* **86**, 367–73.

Rixon, A. J., and Bridge, B. J. (1968). Respiration quotient arising from microbial activity in relation to matric suction and air-filled pore space of soil. *Nature, London* **218**, 961–2.

Robertson, N. F. (1958). Observations of the effect of water on the hyphal apices of *Fusarium oxysporum*. *Annals of Botany*, N.S. **22**, 159–73.

Robertson, N. F., and Rizvi, S. R. H. (1968). Some observations on the water relations of the hyphae of *Neurospora crassa*. *Annals of Botany*, N.S. **32**, 279–91.

Robinson, R. A., and Stokes, R. H. (1955). *Electrolyte Solutions*. New York: Academic Press.

Rose, C. W. (1966). *Agricultural Physics*. Oxford: Pergamon Press.

Rose, D. A. (1968). Water movement in dry soils. I. Physical factors affecting sorption of water in dry soil. *Journal of Soil Science* **19**, 81–93.

Rovira, A. D. (1965*a*). Plant root exudates and their influence upon soil microorganisms. In *Ecology of Soil-Borne Plant Pathogens*, K. F. Baker and W. C. Snyder, eds., pp. 170–86. Berkeley: University of California Press.

Rovira, A. D. (1965*b*). Interactions between plant roots and soil microorganisms. *Annual Review of Microbiology* **19**, 241–66.

Rovira, A. D. (1969). Plant root exudates. *Botanical Review* **35**, 35–58.

Rovira, A. D., and Greacen, E. L. (1957). The effect of aggregate disruption on the activity of microorganisms in the soil. *Australian Journal of Agricultural Research* **8**, 659–73.

Rovira, A. D., and McDougall, B. M. (1967). Microbiological and biochemical aspects of the rhizosphere. In *Soil Biochemistry*, A. D. McLaren and G. F. Peterson, eds., pp. 417–63. New York: Marcel Dekker.

Russell, E. W. (1961). *Soil Conditions and Plant Growth* (9th Edition). London: Longmans.

Russell, E. W. (1968). The agricultural environment of soil bacteria. In *The Ecology of Soil Bacteria*, T. R. G. Gray and D. Parkinson, eds., pp. 77–89. Liverpool: Liverpool University Press.

Ryan, F. J., Beadle, G. W., and Tatum, E. L. (1943). The tube method of measuring the growth rate of *Neurospora*. *American Journal of Botany* **30**, 784–99.

Rybalkina, A. V., and Kononenko, E. V. (1957). Methode d'étude de la microflore active des sols. *Pédologie* **7** (numero spécial), 190–6.

Sabey, B. R. (1969). Influence of soil moisture tension on nitrate accumulation in soils. *Proceedings of the Soil Science Society of America* 33, 263–6.

Saksena, S. B. (1955). Ecological factors governing distribution of soil microfungi in some forest soils of Sagar. *Journal of the Indian Botanical Society* 34, 262–78.

Sasser, J. W., and Jenkins, W. R. (1960). *Nematology*. Chapel Hill: University of Carolina Press.

Scarsbrook, C. E. (1965). Nitrogen availability. In *Soil Nitrogen*, Agronomy Monograph 10, W. V. Bartholomew and F. E. Clark, eds., pp. 481–502. New York: Academic Press.

Schein, R. D. (1964). Comments on the moisture requirements of fungus germination. *Phytopathology* 54, 1427.

Schelhorn, M. von. (1950). Untersuchungen über den Verberb wasserarmer Lebensmittel durch osmophile Mikroorganismen. I. Verberb von Lebensmittel durch osmophile Hefen. *Zeitschrift für Lebensmittel-Untersuchung und-forschung* 91, 117–24.

Schneider, R. (1954). Untersuchungen über Feuchtigkeitsansprüche parasitischer Pilze. *Phytopathologische Zeitschrift* 21, 63–78.

Schroth, M. N., and Cook, R. J. (1964). Seed exudation and its influence on pre-emergence damping-off of bean. *Phytopathology* 54, 670–3.

Schroth, M. N., and Hildebrand, D. C. (1964). Influence of plant exudates on root-infecting fungi. *Annual Review of Phytopathology* 2, 101–32.

Schroth, M. N., Weinhold, A. R., and Hayman, D. S. (1966). The effect of temperature on quantitative differences in exudates from germinating seeds of bean, pea and cotton. *Canadian Journal of Botany* 44, 1429–32.

Schuepp, H., and Frei, E. (1969). Soil fungistasis with respect to pH and profile. *Canadian Journal of Microbiology* 15, 1273–9.

Scott, M. R. (1956a). Studies of the biology of *Sclerotium cepivorum* Berk. I. Growth of mycelium in the soil. *Annals of Applied Biology* 44, 576–83.

Scott, M. R. (1956b). Studies of the biology of *Sclerotium cepivorum* Berk. II. The spread of white rot from plant to plant. *Annals of Applied Biology* 44, 584–9.

Scott, P. R. (1969a). Effects of nitrogen and glucose on saprophytic survival of *Ophiobolus graminis* in buried straw. *Annals of Applied Biology* 63, 27–36.

Scott, P. R. (1969*b*). Control of survival of *Ophiobolus graminis* between consecutive crops of winter wheat. *Annals of Applied Biology* **63**, 37–43.

Scott, W. J. (1953). Water relations of *Staphylococcus aureus* at 30°C. *Australian Journal of Biological Sciences* **6**, 549–64.

Scott, W. J. (1957). Water relations of food spoilage microorganisms. *Advances in Food Research* **7**, 83–127.

Seidel, D. (1965). Untersuchungen über die Keimhemmung von Pilzsporer im Boden mit Hilfe des Agarscheibentests. *Zentralblatt für Bakteriologie, Parasitekunde, Infektionskrankheiten und Hygiene*, Abt. II, **119**, 74–87.

Sewell, G. W. F. (1959*a*). Direct observation of *Verticillium albo-atrum* in soil. *Transactions of the British Mycological Society* **42**, 312–21.

Sewell, G. W. F. (1959*b*). Studies of fungi in a *Calluna*–heathland soil. I. Vertical distribution in soil and on root surfaces. *Transactions of the British Mycological Society* **42**, 343–53.

Sewell, G. W. F. (1959*c*). Studies of fungi in a *Calluna*–heathland soil. II. By the complementary use of several isolation methods. *Transactions of the British Mycological Society* **42**, 354–69.

Shantz, H. L., and Piemeisel, R. L. (1917). Fungus fairy rings in eastern Colorado and their effect on vegetation. *Journal of Agricultural Research* **11**, 191–246.

Siegel, S. M., Roberts, K., Lederman, M., and Daly, O. (1967). Microbiology of saturated salt solutions and other harsh environments. II. Ribonucleotide dependency in the growth of a salt-habituated *Penicillium notatum* in salt-free nutrient media. *Plant Physiology* **42**, 201–4.

Siegenthaler, P. A., Belsky, M. M., and Goldstein, S. (1967). Phosphate uptake in an obligately marine fungus: a specific requirement for sodium. *Science* **155**, 93–94.

Simpson, M. E., and Marsh, P. B. (1969). Microscopic observations on fungi associated with cotton boll-rot fibres. *Mycologia* **61**, 987–96.

Sinha, R. N. (1964). Ecological relationships of stored-products mites and seed-borne fungi. *Acarologia* **6**, 372–89.

Sinha, R. N. (1966). Feeding and reproduction of some stored-product mites on seed-borne fungi. *Journal of Economic Entomology* **59**, 1227–32.

Skujins, J. L., and McLaren, A. D. (1967). Enzyme reaction rates at limited water activities. *Science* **158**, 1569–70.

Slatyer, R. O. (1967). *Plant–Water Relationships*. New York: Academic Press.

Smiles, D. E., and Griffin, D. M. (1966). The measurement of the diffusion of oxygen in saturated porous media. *Australian Journal of Soil Research* 4, 87–93.

Smith, A. M., and Griffin, D. M. (1971). Oxygen and the ecology of *Armillariella elegans* Heim. *Australian Journal of Biological Sciences* 24, 231–62.

Smith, D., Muscatine, L., and Lewis, D. (1969). Carbohydrate movement from autotrophs to heterotrophs in parasitic and mutualistic symbiosis. *Biological Reviews* 44, 17–90.

Smith, R. S. (1964). Effect of diurnal temperature fluctuations on linear growth rate of *Macrophomina phaseoli* in culture. *Phytopathology* 54, 849–52.

Snow, D. (1949). The germination of mold spores at controlled humidities. *Annals of Applied Biology* 36, 1–13.

Somerville, D. A., and Marples, M. J. (1967). The effects of soil enrichment on the isolation of keratinophilic fungi from soil samples. *Sabouraudia* 6, 70–6

Sommers, L. E., Harris, R. F., Dalton, F. N., and Gardner, W. R. (1970). Water potential relations of three root-infecting *Phytophthora* species. *Phytopathology* 60, 932–4.

Spanner, D. C. (1951). The Peltier effect and its use in the measurement of suction pressure. *Journal of Experimental Botany* 2, 145–68.

Stanier, R. Y. (1953). Adaptation, evolutionary and physiological: or Darwinism among the micro-organisms. In *Adaptation in Micro-organisms*, 3rd Symposium of the Society for General Microbiology, R. Davies and E. F. Gale, eds., pp. 1–20. Cambridge: Cambridge University Press.

Steiner, G. W., and Lockwood, J. L. (1969). Soil fungistasis: sensitivity of spores in relation to germination time and size. *Phytopathology* 59, 1084–92.

Steiner, G. W., and Lockwood, J. L. (1970). Soil fungistasis: mechanism in sterilized, re-inoculated soil. *Phytopathology* 60, 89–91.

Stolzy, L. H., and Letey, J. (1964a). Characterizing soil oxygen conditions with a platinum microelectrode. *Advances in Agronomy* 16, 249–79.

Stolzy, L. H., and Letey, J. (1964b). Measurement of oxygen diffusion rates with the platinum microelectrode. III. Correlation of plant response to soil oxygen diffusion rates. *Hilgardia* 35, 567–76.

Stotzky, G. (1966a). Influence of clay minerals on microorganisms. II. Effect of various clay species, homoionic clays, and other particles on bacteria. *Canadian Journal of Microbiology* 12, 831–48.

Stotzky, G. (1966b). Influence of clay minerals on microorganisms. III.

Effect of particle size, cation exchange capacity, and surface area on bacteria. *Canadian Journal of Microbiology* **12**, 1235–46.

Stotzky, G., and Martin, R. T. (1963). Soil mineralogy in relation to the spread of Fusarium wilt of banana in Central America. *Plant and Soil* **18**, 317–38.

Stotzky, G., and Norman, A. G. (1961*a*). Factors limiting microbial activities in soil: I. The level of substrate nitrogen and phosphorus. *Archiv für Mikrobiologie* **40**, 341–69.

Stotzky, G., and Norman, A. G. (1961*b*). Factors limiting microbial activities in soil: II. The effect of sulphur. *Archiv für Mikrobiologie* **40**, 370–82.

Stotzky, G., and Post, A. H. (1967). Soil mineralogy as a possible factor in the geographic distribution of *Histoplasma capsulatum*. *Canadian Journal of Microbiology* **13**, 1–7.

Stotzky, G., and Rem, L. T. (1966). Influence of clay minerals on microorganisms. I. Montmorillonite and kaolinite on bacteria. *Canadian Journal of Microbiology* **12**, 547–63.

Stotzky, G., and Rem, L. T. (1967). Influence of clay minerals on microorganisms. IV. Montmorillonite and kaolinite on fungi. *Canadian Journal of Microbiology* **13**, 1535–50.

Stover, R. H., and Freiberg, S. R. (1958). Effect of carbon dioxide on multiplication of *Fusarium* in soil. *Nature, London* **181**, 788–9.

Strong, D. H., Foster, E. F., and Duncan, C. L. (1970). Influence of water activity on the growth of *Clostridium perfringens*. *Applied Microbiology* **19**, 980–7.

Sussman, A. S. (1966). Dormancy and spore germination. In *The Fungi*, Vol. 2, G. C. Ainsworth and A. S. Sussman, eds., pp. 733–64. New York: Academic Press.

Sussman, A. S. (1968). Longevity and survivability of fungi. In *The Fungi*, Vol. 3, G. C. Ainsworth and A. S. Sussman, eds., pp. 447–86. New York: Academic Press.

Taylor, G. S. (1964). *Fusarium oxysporum* and *Cylindrocarpon radicicola* in relation to their association with plant roots. *Transactions of the British Mycological Society* **47**, 381–92.

Te Strake, D. (1959). Estuarine distribution and saline tolerance of some Saprolegniaceae. *Phyton, Buenos Aires* **12**, 147–52.

Thornton, R. H. (1960). Fungi of some forest and pasture soils. *New Zealand Journal of Agricultural Research* **3**, 699–711.

Toler, R. W., Dukes, P. D., and Jenkins, S. F. (1966). Growth response of *Fusarium oxysporum* f. *tracheiphilum* in vitro to varying oxygen and carbon dioxide tensions. *Phytopathology* **56**, 183–6.

Tomkins, R. G. (1929). Studies of the growth of molds. I. *Proceedings of the Royal Society* B **105**, 375–401.

Toohey, J. I., Nelson, C. D., and Krotkov, G. (1965). Barren ring, a description and study of causal relationships. *Canadian Journal of Botany* **43**, 1043–54.

Tribe, H. T. (1960). Aspects of decomposition of cellulose in Canadian soils. I. Observations with the microscope. *Canadian Journal of Microbiology* **6**, 309–16.

Tribe, H. T. (1966). Interactions of soil fungi on cellulose film. *Transactions of the British Mycological Society* **49**, 457–66.

Tribe, H. T., and Williams, P. A. (1967). Investigations into the basis of microbial ecology in soil, illustrated with reference to growth of soil diphtheroids and *Azotobacter* in a model system. *Canadian Journal of Microbiology* **13**, 467–80.

Trinci, A. P. J. (1969). A kinetic study of the growth of *Aspergillus nidulans* and other fungi. *Journal of General Microbiology* **57**, 11–24.

Trinci, A. P. J., and Whittaker, C. (1968). Self-inhibition of spore germination in *Aspergillus nidulans*. *Transactions of the British Mycological Society* **51**, 594–6.

Trujillo, E. E., and Snyder, W. C. (1963). Uneven distribution of *Fusarium oxysporum* f. *cubense* in Honduras soils. *Phytopathology* **53**, 167–70.

Tsao, P. H. (1970). Selective media for isolation of pathogenic fungi. *Annual Review of Phytopathology* **8**, 157–86.

Tsao, P. H., and Canetta, A. C. (1964). Comparative study of quantitative methods used for estimating the population of *Thielaviopsis basicola* in soil. *Phytopathology* **54**, 633–5.

Turner, G. J. (1967). Snail transmission of species of *Phytophthora* with special reference to foot rot of *Piper nigrum*. *Transactions of the British Mycological Society* **50**, 251–8.

Van den Berg, L., and Lentz, C. P. (1968). The effect of relative humidity and temperature on survival and growth of *Botrytis cinerea* and *Sclerotinia sclerotiorum*. *Canadian Journal of Botany* **46**, 1477–81.

Van Doren, D. M., and Erickson, A. E. (1966). Factors affecting the platinum microelectrode method for measuring the rate of oxygen diffusion through the soil solution. *Soil Science* **102**, 23–8.

Viennot-Bourgin, G. (1964). Interactions entre les champignons parasites telluriques et les autres organismes composants de la rhizosphere. *Annals de l'Institute Pasteur, Paris* **107** (Suppl.), 21–62.

Viets, F. G. (1967). Nutrient availability in relation to soil water. In *Irrigation of Agricultural Lands*, Agronomy Monograph 11, R. M. Hagan, H. W. Haise and T. W. Edminster, eds., pp. 458–71. New York: Academic Press.

Vilain, M. (1963). L'aération du sol. *Annales Agronomiques, Paris* **14**, 967–98.

Waid, J. S. (1962). Influence of oxygen upon growth and respiratory behaviour of fungi decomposing rye-grass roots. *Transactions of the British Mycological Society* **45**, 479–87.

Waksman, S. A. (1959). *The Actinomycetes: I. Nature, Occurrence and Activities*. Baltimore: Williams and Wilkins.

Wallace, H. R. (1963). *The Biology of Plant Parasitic Nematodes*. London: Arnold.

Walter, H. (1924). Plasmaquellung und Wachstum. *Zeitschrift für Botanik* **16**, 353–417.

Warcup, J. H. (1951). The ecology of soil fungi. *Transactions of the British Mycological Society* **34**, 376–99.

Warcup, J. H. (1957). Studies on the occurrence and activity of fungi in a wheat-field soil. *Transactions of the British Mycological Society* **40**, 237–62.

Warcup, J. H. (1965). Growth and reproduction of soil microorganisms in relation to substrate. In *Ecology of Soil-borne Plant Pathogens*, K. F. Baker and W. C. Snyder, eds., pp. 52–67. Berkeley: University of California Press.

Warcup, J. H. (1967). Fungi in soil. In *Soil Biology*, A Burges and F. Raw, eds., pp. 51–110. London: Academic Press.

Warcup, J. H., and Baker, K. F. (1963). Occurrence of dormant ascospores in soil. *Nature, London* **197**, 1317–18.

Ward, E. W. B. (1966). Preliminary studies on the physiology of *Sclerotinia borealis*, a highly psychrophilic fungus. *Canadian Journal of Botany* **44**, 237–46.

Ward, E. W. B., and Henry, A. W. (1961). Comparative response of two saprophytic and two plant parasitic soil fungi to temperature, hydrogen ion concentration, and nutritional factors. *Canadian Journal of Botany* **39**, 65–79.

Wastie, R. L. (1961). Factors affecting competitive saprophytic colonization of the agar plate by various root-infecting fungi. *Transactions of the British Mycological Society* **44**, 145–59.

Wells, J. M., and Uota, M. (1970). Germination and growth of five fungi in low-oxygen and high-carbon dioxide atmospheres. *Phytopathology* 60, 50–3.

Wilkinson, V., and Lucas, R. L. (1969). Influence of herbicides on the competitive ability of fungi to colonize plant tissues. *New Phytologist* 68, 701–8.

Williams, J. B. (1968). Measurement of total and matric suctions of soil water using thermocouple psychrometer and pressure membrane apparatus. *Journal of Applied Ecology* 5, 263–72.

Williams, S. T., and Mayfield, C. I. (1971). Studies on the ecology of actinomycetes in soil. III. The behaviour of neutrophilic streptomycetes in acid soil. *Soil Biology and Biochemistry* 3, 197–208.

Williams, S. T., and Parkinson, D. (1964). Studies on fungi in a podzol. I. Nature and fluctuations of the fungus flora of the mineral horizons. *Journal of Soil Science* 15, 331–41.

Wilson, A. M., and Harris, G. A. (1968). Phosphorylation in crested wheatgrass seed at low water potentials. *Plant Physiology* 43, 61–5.

Winston, P. W., and Bates, D. H. (1960). Saturated solutions for the control of humidity in biological research. *Ecology* 41, 232–7.

Winter, A. G. (1939). Der Einfluss der physiokalischen Bodenstruktur auf den Infektionsverhauf bei der Ophiobolase des Weizens. *Zeitschrift für Pflanzenkrankheiten, Pflenzenpathologie und Pflanzenschutz* 49, 513–59.

Winter, A. G. (1940). Weitere Untersuchungen uber den Einfluss der Bodenstruktur auf die Infektion des Weizens durch *Ophiobolus graminis*. *Zentralblatt für Bakteriologie, Parasitekunde, Infektionskrankheiten und Hygiene*, Abt. II, 101, 364–88.

Wodzinski, R. J., and Frazier, W. C. (1960). Moisture requirements of bacteria. I. Influence of temperature and pH on requirements of *Pseudomonas fluorescens*. *Journal of Bacteriology* 79, 572–8.

Wodzinski, R. J., and Frazier, W. C. (1961a). Moisture requirements of bacteria. II. Influence of temperature, pH, and malate concentration on requirements of *Aerobacter aerogenes*. *Journal of Bacteriology* 81, 353–8.

Wodzinski, R. J., and Frazier, W. C. (1961b). Moisture requirements of bacteria. III. Influence of temperature, pH, and malate and thiamine concentration on requirements of *Lactobacillus viridescens*. *Journal of Bacteriology* 81, 359–65.

Wood-Baker, A. (1955). Effects of oxygen–nitrogen mixtures on the spore germination of mucoraceous moulds. *Transactions of the British Mycological Society* **38**, 291-7.

Wright, J. M. (1956). The production of antibiotics in soil. III. Production of gliotoxin in wheatstraw buried in soil. *Annals of Applied Biology* **44**, 461-6.

Yamaguchi, M., Flocker, W. J., and Howard, F. D. (1967). Soil atmosphere as influenced by temperature and moisture. *Proceedings of the Soil Science Society of America* **31**, 164-7.

Yano, T., Kodama, T., and Yamada, K. (1961). Fundamental studies on the aerobic fermentation. VIII. Oxygen transfer within a mold pellet. *Agricultural and Biological Chemistry* **25**, 580-4.

Yarwood, C. E., and Sylvester, E. S. (1959). The half-life concept of longevity of plant pathogens. *Plant Disease Reporter* **43**, 125-8.

Yong, R. N., and Warkentin, B. P. (1966). *Introduction to Soil Behaviour.* New York: Macmillan Co.

Youngs, E. G. (1965). Water movement in soils. In *The State and Movement of Water in Living Organisms*, 19th Symposium of the Society for Experimental Biology, C. E. Fogg, ed., pp. 89-112. Cambridge: Cambridge University Press.

Zak, B. (1964). Role of mycorrhizae in root disease. *Annual Review of Phytopathology* **2**, 377-92.

Zalokar, M. (1965). Integration of cellular metabolism. In *The Fungi*, Vol. 1, G. C. Ainsworth and A. S. Sussman, eds., pp. 377-428. New York: Academic Press.

ZoBell, C. (1943). The effect of solid surfaces upon bacterial activity. *Journal of Bacteriology* **46**, 39-56.

Zvyagintsev, D. G., Pertsovskaya, A. F., Duda, V. I., and Nikitin, D. I. (1969). Electron microscope study of adsorption of microorganisms on soils and minerals. *Mikrobiologiya* **38**, 1091-5.

# Index